Superconductivity in Science and Technology

Edited by Morrel H. Cohen

Superconductivity in Science and Technology

University of Chicago Press
Chicago & London

Library of Congress Catalog Card Number: 67-25534
The University of Chicago Press, Chicago 60637
The University of Chicago Press, Ltd., London W. C. 1
© 1968 by The University of Chicago. All rights reserved
Published 1968. Printed in the United States of America

Preface

Before the last decade, research in superconductivity was the province of a relatively small and devoted group of investigators working within the larger fields of solid state and low temperature physics. The advent in 1957 of the Bardeen–Cooper–Schrieffer theory, the first successful microscopic theory of superconductivity, led immediately to a rapidly accelerating tempo of research on, and a broadening of interest in, superconductivity. The last ten years have seen a flow of stimulation outward from superconductivity to particle physics, nuclear physics, astrophysics, and quantum statistical mechanics, and a flow of interest inward from metallurgy, chemistry, and electrical engineering as well as from other areas of solid state physics.

The absence of electrical resistance in superconductors has always held promise of important technological applications since its discovery by Kamerlingh–Ohnes in 1911. The activity of the last decade, together with concomitant technological advances, has brought that promise close to realization. High field, superconducting magnets are now commercially available; various superconducting electronic devices have been invented; and applications to computers and to electric power transmission, *inter alia*, are under active discussion. Accordingly, a conference on "Superconductivity in Science and Technology" was held at the University of Chicago on May 23 and 24, 1966, during the Seventy-Fifth Anniversary Year of the University, as a part of the Industrial Sponsors Program in Basic Research in the Physical Sciences at the University of Chicago. A group of recognized experts on both scientific and technical aspects of superconductivity were brought together in the Conference to provide the background for partial answers to the question whether we can expect large scale technological applications of superconductivity, and if so, when. The papers formally presented at the Conference appear here as the seven chapters of this book.

The first three chapters are concerned with the scientific aspects of superconductivity. In Chapter 1, Bardeen introduces the entire subject, discussing the theory in simple terms and pointing towards the unsolved problems. Josephson discusses superconductive tunneling in Chapter 2, again with primary emphasis on the physical phenomena and their interpretation. The characterization and classification of superconducting materials is taken up by Berlincourt in Chapter 3, a complete, well-organized, and well-documented review of the subject. Together, these first three chapters introduce the background physics necessary for an appreciation of the remaining four chapters devoted to the technological aspects of superconductivity, while Chapter 3 by itself is an important contribution to the scientific literature on superconductivity.

Chapters 4 and 5 deal with specific technological applications of superconductivity. Mercereau discusses instruments that exploit the macroscopic quantal character of superconductivity, in Chapter 4. Garwin and Matisoo address themselves in Chapter 5 to what appears to be the single largest technological question for superconductivity, the feasibility of superconducting power transmission lines. Chapters 6 and 7, on the other hand, are broader gauged. Hulm, in Chapter 6, reviews the outlook for superconductivity in technology in terms of past proposals and future prospects. Schmitt and Morrison treat the economic aspects of superconductive technology in some detail in Chapter 7, which might be subtitled "What we have spent, and is it worth it?"

With the exception of the chapter by Garwin and Matisoo (1967, *Proc. IEEE* 55: 538), none of the material of this book has been published elsewhere. The Conference proved of interest to those actively engaged in scientific research and technological development relating to superconductivity, to corporate executives and government officers responsible for long-range planning within their institutions, and to those more generally concerned with the interplay of science and technology. It is hoped that the publication of this book will bring the substance of the Conference to a wider audience of similar composition.

I am deeply grateful to Mrs. Ruth L. Patterson, David M. Huntington, and Professor David H. Douglass for their assistance in organizing and running the Conference, and to Miss Judith Ruby and K. L. Ngai for the construction of the indexes. We are also grateful to the IEEE for permission to reprint the Garwin–Matisoo paper, and to the *Physical Review* and the *Journal of Experimental and Theoretical Physics* (USSR) for permission to reprint various of the figures.   MORREL H. COHEN

Contents

1	Theory of Superconductivity John Bardeen	1
2	Superconductive Tunneling Brian D. Josephson	19
3	Superconducting Materials Ted G. Berlincourt	31
4	Quantum Engineering James E. Mercereau	63
5	Superconducting Lines for the Transmission of Large Amounts of Electrical Power over Great Distances Richard L. Garwin and J. Matisoo	77
6	Superconductors in Technology John K. Hulm	103
7	Economic Aspects of Superconductivity Roland W. Schmitt and W. Adair Morrison	127
	Subject Index	157
	Name Index	161

1
Theory of Superconductivity
John Bardeen

Introduction

Superconductivity is connected with a second-order phase transition of the electrons arising from pairing of electrons in the ground state. An important concept is a condensate wave function that describes the pairing and a supercurrent flow that can vary in space and time. The condensate wave function is closely related to the energy-gap parameter, $\Delta(\mathbf{r}, t)$, which is a complex function of position and time. More generally, Δ should be a function of two position variables, $\mathbf{r}_1$ and $\mathbf{r}_2$, representing the coordinates of the pair, but one may regard it as depending only on the average position $\mathbf{r} = \tfrac{1}{2}(\mathbf{r}_1 + \mathbf{r}_2)$ provided that the space variations are sufficiently slow.

From a macroscopic point of view, supercurrent flow can be described at least approximately in terms of an effective wave function of this sort with amplitude and phase. This leads to a description of a superconductor as a quantum system on a macroscopic scale, as suggested by the late Fritz London [1] many years ago, and which is inherent in the phenomenological equations of Ginzburg and Landau [2]. Many beautiful experiments, including demonstrations of flux quantization and of quantum interference effects in Josephson tunnel junctions, show in striking ways the quantum aspects.

To help put things in perspective, I plan to first give an outline of some of the important experimental and theoretical developments that

---

The author is professor in the Department of Physics and Materials Research Laboratory, University of Illinois, Urbana, Illinois.

have led to our present understanding. Next, I will discuss the nature of the quasi-particle excitation spectrum in normal and superconducting metals. The many-particle wave function describing the ground state of a superconductor may be regarded as a linear combination of low-lying normal configurations in which the quasi-particle states of individual electrons are occupied in pairs of opposite spin and momentum. This leads to the introduction of the gap parameter, Δ, and to a set of quasi-particle excitations in a superconductor in one-to-one correspondence with those of the normal metal. From a knowledge of the excitation spectrum of a superconductor, one can calculate many equilibrium and transport properties of superconductors without a great deal more difficulty than for normal metals. One example, the calculation of the surface impedance at microwave frequencies, will be discussed. Following that, I will show how supercurrent flow is related to the gradient of the phase of the condensate wave function, and give the application to Abrikosov vortex lines in Type II superconductors. Finally, I will discuss some of the outstanding problems of superconductivity theory.

Historical Outline

Listed below are the dates of some of the more important discoveries concerning superconductivity.

1911—Discovery, H. K. Onnes [3]. Persistent currents, critical fields.

1933—Flux exclusion, Meissner and Ochsenfeld [4].

1935—London phenomenological theory and F. London's proposal that superconductivity is a quantum phenomenon [5].

1950—Isotope effect, $T_c \sim M^{-1/2}$, Reynolds et al. [6]; Maxwell [7]. Importance of electron–phonon interactions; Fröhlich [8].

1953—Experimental evidence for an energy gap. Goodman [9], others.

1957—Microscopic theory based on pairing. Bardeen, Cooper, and Schrieffer [10].

1960—Tunnel effect. Giaever [11].

1961—Flux quantization. Deaver and Fairbank [12]; Doll and Näbauer [13].

1962—Supercurrent flow through tunnel barrier. Josephson [14].

Superconductivity was discovered by Kamerling Onnes [3] in 1911, not long after he learned to liquefy helium. One of his first ideas was to make a superconducting magnet, but he soon found that superconductivity is destroyed by a modest magnetic field of a few hundred gauss. Only in the past few years have materials been found that remain superconducting at very high fields; these are the Type II superconductors we shall discuss later.

Until 1933, superconductivity was thought to be simply a case of vanishing electrical resistance; it was thought that electrons could somehow move through a superconductor without being scattered. In that year, Meissner and Ochsenfeld [4] discovered another aspect perhaps even more basic: a superconductor is a perfect diamagnet as well as a perfect conductor. A superconductor excludes a magnetic field except in a small penetration region near the surface where currents flow so as to balance the applied field and make the total field vanish in the interior. The thickness of this penetration layer, the so-called penetration depth, is generally about 500 Å.

If a high-frequency alternating field is applied, the time rate of change of magnetic field in the penetration region gives rise to an electric field; this field can accelerate the electrons near the surface and give rise to dissipation. Thus a superconductor has zero loss only under dc conditions. However, in Type I superconductors the loss is appreciable only at very high frequencies. It is possible to make cavities with superconducting walls that have extremely high Q at microwave frequencies.

Not long after the discovery of the Meissner effect, the London brothers [5] proposed their phenomenological theory to describe both infinite conductivity and diamagnetic aspects. Not long after, Fritz London suggested that the explanation of superconductivity was to be found as a quantum phenomenon on a macroscopic scale; his general view of the nature of superconductivity has been substantiated by present microscopic theory. Nevertheless, the development of a satisfactory microscopic theory remained a puzzle for a long time.

An important breakthrough was the experimental discovery of the isotope effect [6, 7], and the independent suggestion of Fröhlich [8] that superconductivity is somehow connected with the interaction between electrons and lattice vibrations. He pointed out that the interaction between the electrons and the lattice vibrations or phonons gives an effective attractive interaction between the electrons. Although this gave an important clue, it still remained a problem to construct an adequate microscopic theory.

Starting about 1953, a number of experiments were carried out that indicated an energy gap in the quasi-particle excitation spectrum of superconductors. This gap is one of the important features in the present microscopic theory. In 1957, Cooper, Schrieffer, and I [10] proposed a microscopic theory based on pairing of electrons of opposite spin and momentum. The theory has since then been extended and amplified in many directions, so that it could be applied to a wider variety of problems. The theory has been used to predict new phenomena which have been later demonstrated experimentally.

One of the most important of these discoveries is that of tunneling of electrons through a thin insulating layer between a superconductor and a normal metal or between two superconductors. This discovery was made by Giaever [11] about 1960; he thought that tunneling might be a good way to demonstrate the energy gap in a superconductor and carried out the experiment. Many beautiful experiments have grown out of this very important discovery.

A couple of years later, Josephson [14] suggested that if two superconductors are separated by a very thin insulating layer a supercurrent can tunnel between them. In other words, current can flow with no voltage applied between the two superconductors. This current flow is very sensitive to magnetic fields. Josephson tunneling was first demonstrated experimentally by Rowell and Anderson [15] and since has been studied by many others. Since Josephson discusses this topic in a following chapter, little will be said about it here.

I would like to go back in time to discuss briefly the history of Type II superconductivity. The phenomenon was discovered experimentally by a Russian physicist, Schubnikov [16], around 1937. He found that some superconductors seem to allow flux penetration. When the applied magnetic field is above a lower critical field, called H_{c1}, flux begins to penetrate but the material remains superconducting until a higher critical field, the upper critical field, H_{c2}, is reached. With a perfect Meissner effect flux is completely excluded. A plot of the magnetic moment of a long cylinder as a function of an applied magnetic field along the axis is a straight line until there is a transition from the superconducting state to the normal state at the thermodynamic critical field, H_c. Schubnikov found that in some superconductors, called Type II, the magnetic moment starts to decrease at a field $H_{c1} < H_c$, but superconductivity remains until the upper critical field $H_{c2} > H_c$ is reached. Recently, predictions of Saint James and De Gennes [17] and subsequent experiments show that there is a sheath near the surface of the superconductor that remains superconducting to a still higher critical field, H_{c3}.

In Type II superconductors, flux penetrates by means of quantized vortex lines about which electrons circulate and give rise to a magnetic field. To describe such circulating supercurrents, it is necessary to have a theory that allows for space variations of supercurrent flow. Such a theory was given on phenomenological grounds by Ginzburg and Landau [2] about 1950. On the basis of this theory, Abrikosov [18] in 1953 worked out his now famous theory of Type II superconductors based on flux penetration in the form of an array of quantized vortex lines. This work was not published until 1957. The Ginzburg–Landau theory has been shown by Gor'kov [19] to follow from the microscopic theory in certain limiting cases, particularly near the transition temperature T_c. According to present microscopic theory, modifications are required at other temperatures, but Abrikosov's theory and its subsequent developments form the basis of our present understanding of Type II superconductors. It is known as the GLAG theory (Ginzburg, Laudau, Abrikosov, Gor'kov).

Microscopic Theory

Next, I would like to discuss briefly the basis for the present microscopic theory of superconductivity [10]. We first review the electronic structure of normal metals. One can describe the various states of individual electrons (Bloch states) in normal metals in terms of a wave vector **k** or an associated momentum $\mathbf{p} = \hbar\mathbf{k}$. In a free electron model the energy $\mathscr{E}$ as a function of momentum is just given by $\mathscr{E} = \mathbf{p}^2/2m$. In an actual metal in which the electrons move in the periodic field of the ions, the energy of a Bloch electron is still a function of the momentum but may differ from the free electron value. Results of energy band theory may be used to describe the excitation spectrum of normal metals. While we talk about individual electrons, actually there are, of course, strong interactions between them from Coulomb forces and from electron-phonon interactions. Nevertheless, the excited states of a normal metal can be described pretty well in terms of the Bloch one-particle picture. Various excited states of the metal can be described by giving the occupation of Bloch states above the Fermi level and the unoccupied states or holes below the Fermi level. At the absolute zero, the lowest energy states consistent with the exclusion principle are occupied; this means that all of the states up to the Fermi surface of energy $\mathscr{E}_F$, momentum p_F, are occupied and the states outside are unoccupied. There is a large number of low-lying excited configurations in which electrons are excited above the Fermi surface with a corresponding number of unoccupied states or holes below. Figure 1 illustrates a typical excited configuration of a normal metal. It is convenient to express the energies of individual electrons as measured from the Fermi energy, $\epsilon_p = \mathscr{E}_p - \mathscr{E}_F$. Then the energy of an excited configuration is obtained by adding the excitation energies of all the excited electrons and holes, taking the energy of a hole to be $|\epsilon_p|$ for $p < p_F$; this gives the excitation energy relative to the Fermi sea.

In quantum theory one can form a wave function by taking linear combinations of other wave functions. We think of the ground state of a superconductor as formed from a linear combination of normal configurations of a particular sort designed to take advantage of the effective attractive interaction between the electrons. There are matrix elements of the interactions between different configurations in which a pair of electrons is scattered from one set of states to another set consistent with overall conservation of momentum. One can get a lowering of energy if the configurations are chosen in such a way that the matrix elements of the interaction between configurations are predominantly negative. This determines that the configurations which enter the superconducting ground state are those in which the individual electron states are occupied in pairs of opposite spin and momentum. Thus if momentum **p** and spin up is occupied, its partner of opposite spin and opposite momentum is also occupied. There will be some configurations in which both of these are occupied and other configurations in which both of these are empty, but in the ground state there will be no configurations in which only one state of the pair is occupied and the other is empty.

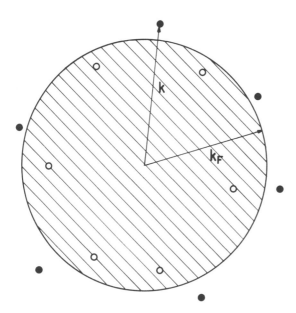

Fig. 1 Schematic diagram of a typical configuration. In the configurations included in the ground state wave function of a superconductor, the quasi-particle states are occupied in pairs of opposite spin and momentum.

One can construct excited states of a superconductor in which electrons, or quasi-particles, are excited out of the ground state. These excited configurations are in one-to-one correspondence with those of the normal metal. One such quasi-particle excited state would correspond to a state of momentum **p** and spin up occupied but its partner with spin down unoccupied in all configurations, and with other momentum states being occupied in pairs. Relative to the ground state, the quasi-particle energy E_p in a superconductor is

$$E_p = \sqrt{\epsilon_p^2 + \Delta^2}, \qquad (1)$$

where ϵ_p is the energy of the corresponding state in the normal metal and Δ is the energy gap parameter. The minimum energy required to create a pair of excitations, corresponding to an electron and a hole, from the ground state is 2Δ.

A more basic definition of Δ is in terms of the wave function describing the ground state. There is a certain probability of a given pair being occupied, and a certain probability that it is unoccupied. The probability that it is occupied is

$$|u_p|^2 = \frac{1}{2}\left(1 - \frac{\epsilon_p}{E_p}\right). \qquad (2)$$

When the energy is well above the Fermi surface ϵ_p is large and positive, $|u_p|^2 \to 0$. Well below the Fermi surface, where ϵ_p is large and negative, $|u_p|^2 \to 1$. Thus the energy gap parameter describes the nature of the

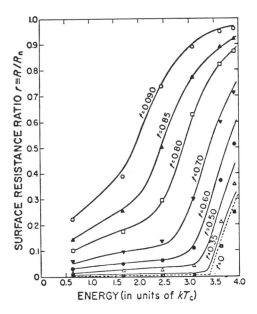

Fig. 2 Frequency dependence of the surface resistance of aluminum at various reduced temperatures $t = T/T_c$, as measured by Biondi and Garfunkel [20] (smooth curves). Points are from calculations of P. B. Miller [21] based on the microscopic theory.

pairing in the ground state, and this is its more general significance. In the simplified version of the theory we have discussed, but not in some of its extensions, Δ also gives the quasi-particle excitation spectrum.

Given a complete set of excitations, with energies and wave functions, one can calculate various transport properties such as the response to electromagnetic fields, thermal conductivity, ultrasonic attenuation, spin relaxation times, and so on. This has been generally quite successful, with excellent agreement between theory and experiment.

I shall just give one example, a calculation of the surface resistance of aluminum as a function of frequency. Figure 2 is a plot of the surface resistance ratio, the ratio of the surface resistance, R_s, in the superconducting state to that in the normal state, R_n. Quantum energies, $h\nu$, are expressed in units of kT_c. The experimental data are those of Biondi and Garfunkel [20]. Opposite to the usual way of plotting, the solid curves represent the experimental data, and the plotted points theoretical values. The latter are from calculations of a former student, P. B. Miller [21].

There is excellent agreement between theory and experiment. At very low temperatures there is no absorption until $h\nu > 2\Delta$. At higher temperatures, quasi-particle excitations present in thermal equilibrium can be scattered, giving rise to absorption of energy. The knee of the curve corresponds to the increased absorption that occurs when $h\nu > 2\Delta(T)$. The energy gap, $2\Delta(T)$, decreases with increasing temperature, going to zero at T_c.

To have low loss at microwave frequencies at temperatures of $\sim 2°K$, it is necessary to use superconductors with a relatively high transition temperature, such as lead for which $T_c \simeq 7.2°K$.

Superfluid Flow

So far we have been talking about the ground state in which there is no current flow and about quasi-particle excitations from this ground state. For many applications, one is interested in describing supercurrent flow in which the current density may vary in space and time. In an infinite system, one can get a state with current flow by displacing the ground state in momentum space by $m\mathbf{v}_s$. There is still pairing in that an electron of momentum $\mathbf{p} + m\mathbf{v}_s$ with spin up is paired with an electron of momentum $-\mathbf{p} + m\mathbf{v}_s$, spin down, where $\mathbf{v}_s$ is the velocity of the supercurrent flow. The total momentum is identical for all pairs, $2m\mathbf{v}_s$. Thus there is macroscopic occupation of this momentum state in pairs.

Macroscopic occupation is the key to the understanding of the superfluid properties, both in superconductors and in superfluid helium. In the latter, macroscopic occupation comes from the Einstein–Bose condensation which persists even when interactions between the atoms in the liquid are taken into account. Many like to think of a superconductor as an Einstein–Bose condensation of bound pairs of electrons. There is some analogy, but there are also marked differences since the electrons obey Fermi–Dirac statistics. The momentum state of macroscopic occupation defines a unique reference frame for both systems. Starting from rest, if one displaces the system by a velocity $\mathbf{v}_s$, the total flow is given by $\rho\mathbf{v}_s$, where ρ is the density. In a normal metal, scattering of electrons soon reduces the current to zero. A superconductor differs in that after scattering takes place and a steady state is reached, a net flow, $\rho_s\mathbf{v}_s$, remains. In other words, if the ground state pairs are moving with a velocity $\mathbf{v}_s$ and the quasi-particles come into thermal equilibrium appropriate to this value of $\mathbf{v}_s$, there remains a net flow. A superfluid is characterized by a value of ρ_s different from zero. Supercurrent flow is thus a local equilibrium flow that occurs in the presence of scattering. The only way one can change this flow is to change $\mathbf{v}_s$, the common velocity of the ground state pairs. This requires a force acting on all or a large fraction of the electron. Scattering of quasi-particles does not do this, so that the current persists in time.

London [1] showed how one can construct a wave function for a system in which $\mathbf{v}_s$ varies slowly from point to point. If $\Psi_0(\mathbf{r}_1, \mathbf{r}_2 \ldots \mathbf{r}_N)$ is the ground state wave function with no current flow, one can take

$$\Psi(\mathbf{r}_1, \mathbf{r}_2, \ldots \mathbf{r}_N) = \exp\left[i\sum_j \chi(\mathbf{r}_j)\right]\Psi_0(\mathbf{r}_1, \mathbf{r}_2 \ldots \mathbf{r}_N). \tag{3}$$

The local momentum is proportional to the gradient of the phase:

$$\mathbf{p}_s = \hbar \operatorname{grad} \chi. \tag{4}$$

This implies potential flow, curl $\mathbf{p}_s = 0$. The phase $\chi(\mathbf{r})$ corresponds to that of the condensate wave function.

As defined by equation (4), $\mathbf{p}_s$ is the canonical momentum. In the presence of a magnetic field defined by a vector potential, $\mathbf{A}(\mathbf{r})$, the kinetic momentum is

$$\mathbf{P} = m\mathbf{v}_s = \mathbf{p}_s - \frac{e}{c}\mathbf{A}. \tag{5}$$

London [1] showed that as a consequence of these considerations and the fact that $\chi(\mathbf{r})$ must be single-valued, the flux threading a hole in a multiply connected superconductor must be quantized. The line integral of $\mathbf{p}_s$ about any closed path must be a multiple of h:

$$\oint \mathbf{p}_s \cdot d\mathbf{l} = nh. \tag{6}$$

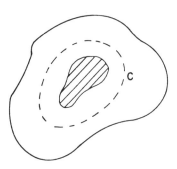

Fig. 3 Path integration encircling a hole in a superconductor.

If, as illustrated in Figure 3, one takes a closed path in the interior of the superconductor where $\mathbf{P} = m\mathbf{v}_s = 0$, then $\mathbf{p}_s = (e/c)\mathbf{A}$. This implies that the flux enclosed by the path,

$$\Phi = \oint \mathbf{A} \cdot d\mathbf{l} = n(hc/e), \qquad (7)$$

is quantized in units of hc/e. The experiments of Deaver and Fairbank [12] and of Doll and Näbauer [13] indicated flux quanta of just half the London value, or $hc/2e \sim 2 \times 10^{-7}$ gauss cm². It was soon shown that this is to be expected as a result of pairing in the ground state. One may regard $\mathbf{P}$ and $\mathbf{p}_s$ as referring to the momenta of a pair, so that

$$\mathbf{P} = 2m\mathbf{v}_s = \mathbf{p}_s - \frac{2e}{c}\mathbf{A}, \qquad (8)$$

which leads to $\Phi = n(hc/2e)$.

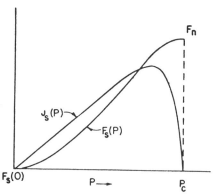

Fig. 4 Schematic plot of free energy, F_s, and of supercurrent density, J_s, as functions of the kinetic momentum, P.

It is only for small values of **P** that the current density is proportional to **P**. There is a very general relation between the current density $\mathbf{J}_s(\mathbf{P})$ and the free energy $F_s(\mathbf{P})$. If $\mathbf{J}_s(\mathbf{P})$ refers to particle rather than electrical current,

$$\mathbf{J}_s(\mathbf{P}) = \partial F_s(\mathbf{P})/\partial \mathbf{P}. \tag{9}$$

This expression applies at finite temperatures and to alloys as well as to pure metals. In calculating the free energy, the state of macroscopic occupation defined by **P** is specified. The quasi-particle excitations are presumed to be in equilibrium for the specified **P**.

A typical plot of J_s and F as a function of P is given in Figure 4. When P is small,

$$F_s(P) - F_s(0) = P^2/4m = \tfrac{1}{2}\rho_s v_s^2. \tag{10}$$

As $F_s(P)$ approaches the free energy in the normal state, F_n, $J_s(P)$ drops to zero and there is a transition to the normal state. The critical value of the momentum where this occurs is called P_c. The supercurrent density is a maximum for a smaller value of P.

The phase plays the role for superfluid flow that voltage does for ordinary flow in normal metals. In a normal metal the current density is proportional to the voltage gradient, in a superfluid to the gradient in phase. If there is no supercurrent flow over "a mile of dirty lead wire," this means that the phase is the same everywhere in the wire.

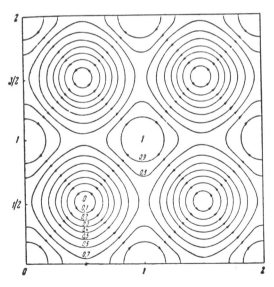

Fig. 5 Current paths in planes normal to the magnetic field in a vortex array in Type II superconductor in the mixed state. (After Abrikosov [18].)

Type II Superconductors

In superconductors of the second kind, when the applied field is above the lower critical field, flux enters in the form of an array of quantized vortex lines. This is called the mixed state. Currents circulate about the axis of a vortex line, giving a magnetic field along its length. The flux associated with a given line is quantized in units of $hc/2e$. It can be shown that for a given total flux density, it is energetically most favorable to have the flux distributed in vortex lines each of which has a single quantum of flux. The current paths associated with such an array are shown in Figure 5, taken from Abrikosov's original paper. His picture has been strikingly confirmed by many experiments, perhaps most directly by the beautiful neutron diffraction experiments of the Saclay group [22].

The amplitude of the condensate wave function vanishes on the axis. Caroli, De Gennes and Matricon [23] have shown that the density of quasi-particle states in the region of the core of the line is approximately that of a normal region of radius equal to the coherence distance. From the viewpoint of the local model we have discussed earlier, the canonical momentum of a pair in the vicinity of the axis is

$$p_{s\theta} = \hbar/r, \qquad (11)$$

where r is the distance from the axis. As r decreases, the kinetic momentum, P, increases, until P reaches P_c, where the metal becomes normal. This determines the radius of the normal core as

$$a = \hbar/P_c \qquad (12)$$

A schematic diagram of an experiment to observe motion of vortex lines is shown in Figure 6. There is a magnetic field normal to a slab of a Type II superconductor, creating an array of vortex lines. A transport current J_T running along the length of the slab produces a force on the lines tending to make them move in the vertical or y-direction. One may regard this force either as a hydrodynamic force or as the usual Ampere's law force between the transport current and the circulating currents of the vortex lines. Motion of the lines in the y-direction gives rise to an electric field in the x-direction corresponding to ordinary resistance. Motion in the x-direction parallel to the transport current gives a Hall field in the y-direction. The average electric field is just what one would calculate as generated from the moving flux lines by induction, with each line carrying a flux unit, $hc/2e$.

A typical experiment for measuring resistive effects in Type II superconductors is shown in Figure 7, taken from a paper of Kim and coworkers [24]. Probes are used to measure longitudinal and transverse

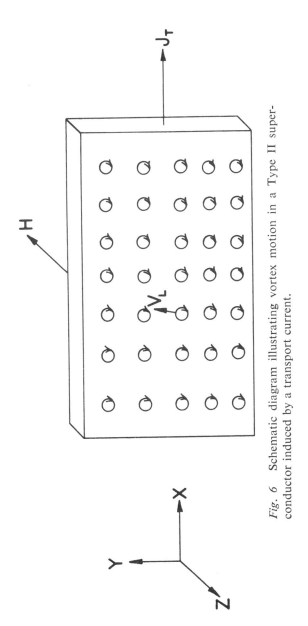

Fig. 6. Schematic diagram illustrating vortex motion in a Type II superconductor induced by a transport current.

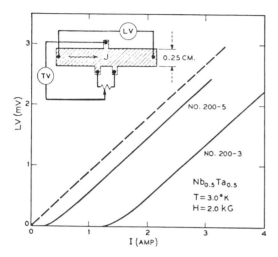

Fig. 7 Measurement of flux flow resistivity of a NbTa alloy in the mixed state. Different specimens of similar composition have different critical currents for the onset of resistivity but the slopes in the linear regions are the same. They are presumed to be the same as for ideal material with no pinning forces. (After Kim, *et al.* [24].)

voltages generated by a transport current. The figure shows the longitudinal voltage measured in a niobium–tantalum alloy as a function of the transport current, I. Ordinarily, the vortex lines are pinned at imperfections and do not move, so that there is no dissipation. If the transport current is increased above a critical value dependent on the particular specimen, the vortex lines become unpinned and a voltage varying linearly with the current is observed. The slope is the same for different specimens measured under similar conditions and is taken to give the ideal flux flow resistivity, ρ_f.

Kim and coworkers found that the resistivity of various metals and alloys studied follows an approximate law of corresponding states,

$$\rho_f = \rho_n \frac{H}{H_{c2}}, \qquad (13)$$

where ρ_n is the normal resistivity and H_{c2} is the upper critical field at the corresponding temperature. One may interpret H/H_{c2} as giving roughly the fraction of the volume occupied by the normal cores. It was noted that the dissipation is about what one would expect if the transport current flowed directly through the normal cores of the vortex lines. Rosenblum and Cardona had earlier proposed such a picture in their interpretation of data on the microwave surface resistance of Type II superconductors. A theory based on the local model gives just this result.

There have been several experiments that indicate that the vortex lines actually do move in the resistive state. Perhaps most direct is that of Giaever [25], who studied the voltage induced in a secondary film separated by a thin insulating layer from a primary film. A vortex pattern penetrating both films was produced by a magnetic field perpendicular to the plane of the films. The spacing was so close that each vortex line threads both films. Motion induced by a transport current in the primary film causes motion of the corresponding lines in the secondary film, and thus generates a voltage in the secondary. In this way, Giaever was able to make a dc transformer. Another investigation indicating motion is the measurement by Van Ooyen and Van Gurp [26] of the noise spectrum of the dc voltage in a flux flow experiment. It was found that the cut-off frequency is determined by the transit time of a moving flux line across the slab. The magnitude of flux in the moving entities varied with the transport current, decreasing from bundles of $\sim 10^5$ flux quanta to ~ 10 flux quanta with increasing current. Other evidence comes from measurements by Fiory and Serin [27] of thermoelectric and by Otter and Solomon [28] of thermomagnetic effects. These experiments indicate that a large heat is transported by the moving vortex lines, and can be interpreted by assuming that the quasi-particle excitations in the cores move with the lines. Theories of these effects given by Stephen [29] are in good accord with experiment.

A theory of flux motion based on the local model as developed in collaboration with Stephen [30] leads to Kim's empirical law, equation (13). A somewhat similar theory has been given independently by van Viejfeijken and Niessen [31]. Electric fields generated by the motion of a vortex line drive the transport current through the normal core. There is dissipation in the transition region just outside the core as well as in the core itself. The theory indicates that the Hall angle in the mixed state should be the same as that in the normal core.

Some Current Problems

There is considerable interest in developing a more adequate theory of vortex motion. Present theory is based on a local model that assumes that changes in the condensate occur slowly in space and time. This is not true in the vicinity of the core where most of the dissipation occurs. There are other problems as well that involve space or time variations of the order parameter, such as the theory of boundary effects.

Another area of theoretical interest is the interaction between magnetic and superconducting properties. What is the relation between ferromagnetism or antiferromagnetism and superconductivity?

One of the most important problems from a practical point of view, and one with which theorists are becoming more and more concerned, is to try to understand the superconducting properties of particular materials. Much of the past work has been concerned with general theories based on idealized models. If one were better able to calculate the properties of particular metals or alloys, one might be able to design materials with desired characteristics. We do not yet understand what determines the upper limit on critical temperatures. Empirically, a number of alloys have been found with values of T_c around 18°K, but none higher. [Editor's note: One with a T_c of 20.05°K has subsequently been found [32].]

These are just a few of the directions of current interest in the theory of superconductivity. It is still a very active field with many unsolved problems.

References

[1] London, F., *Superfluids*, Vol. I. New York: Wiley.
[2] Ginzburg, V. L., and Landau, L. D. 1950. *Zh. Eksperim. i Teor. Fiz.* 20: 1064.
[3] Onnes, H. Kamerlingh 1911. *Comm. Phys. Lab. Univ. Leiden*, Nos. 119, 120, 122.
[4] Meissner, W., and Ochsenfeld, R. 1933. *Naturwiss.* 21: 787; Gorter, C. J. 1933. *Nature* 132: 931.
[5] London, F. 1935. *Proc. Roy. Soc. (London)* A152: 24.
[6] Reynolds, Serin, Wright and Nesbitt. 1950. *Phys. Rev.* 78: 487.
[7] Maxwell, E. 1950. *Phys. Rev.* 78: 477.
[8] Fröhlich, H. 1950. *Phys. Rev.* 79: 845.
[9] Goodman, B. B. 1958. *Proc. Phys. Soc. (London)* A66: 217.
[10] Bardeen, J., Cooper, L. N., and Schrieffer, J. R. 1957. *Phys. Rev.* 108: 1175.
[11] Giaever, I. 1960. *Phys. Rev. Letters* 5: 147.
[12] Deaver, B. S., Jr., and Fairbank, W. M. 1961. *Phys. Rev. Letters* 7: 43.
[13] Doll, R., and Näbauer, M. 1961. *Phys. Rev. Letters* 7: 51.
[14] Josephson, B. D. 1962. *Phys. Letters* 1: 251.
[15] Anderson, P. W., and Rowell, J. M. 1963. *Phys. Rev. Letters* 10: 230.
[16] Schubnikov, L. V., et al. 1937. *Zh. Eksperim. i Teor. Fiz.* 7: 221.
[17] Saint-James, D., and de Gennes, P. G. 1963. *Phys. Letters* 7: 306.
[18] Abrikosov, A. A. 1957. *Zh. Eksperim. i Teor. Fiz.* 32: 1442 [Translation: *Soviet Phys. JETP* 5: 1174].
[19] Gor'kov, L. P. 1959. *Zh. Eksperim. i Teor. Fiz.* 36: 1918 [Translation: *Soviet Phys. JETP* 9: 1364].

[20] Biondi, M. A., and Garfunkel, M. P. 1959. *Phys. Rev.* 116: 853.
[21] Miller, P. B. 1960. *Phys. Rev.* 118: 928.
[22] Cribier, D., Jacrot, B., Rao, L. M., and Farnoux, B. 1964. *Phys. Letters*, 9: 106.
[23] Caroli, C., de Gennes, P. G., and Matricon, J. 1964. *Phys. Letters* 9: 307.
[24] Kim, Y. B., Hempstead, C. F., and Strnad, A. R. 1965. *Phys. Rev.* 139: A1163.
[25] Giaever, I. 1966. *Phys. Rev. Letters* 16: 460.
[26] Van Ooyen, D. J., and van Gurp, G. J. 1965. *Phys. Letters* 17: 230.
[27] Fiory, A. T., and Serin, B. 1966. *Phys. Rev. Letters* 16: 308.
[28] Otter, F. A., and Solomon, P. R. 1966. *Phys. Rev. Letters* 16: 681.
[29] Stephen, M. J. 1966. *Phys. Rev. Letters* 16: 801.
[30] Bardeen, J., and Stephen, M. J. 1965. *Phys. Rev.* 140: A1197. (Note some differences in notation; e.g., there, p_s is the momentum per particle rather than per pair.)
[31] Van Veijfeijken, A. G., and Niessen, A. K. 1965. *Phys. Letters* 16: 23; and *Philips Res. Repts* 20: 205.
[32] Matthias, B. T. *et al.* 1967. *Science* 156: 645.

2
Superconductive Tunneling

Brian D. Josephson

The first experiments on superconductive tunneling were carried out by Giaever in 1960 [1]. As you know, the idea is to make a sandwich in which on the outside you have two pieces of metal, while the jam in between is a very thin layer of insulator called the barrier, which is something like 20 Å thick. It is so thin that it is possible for electrons to tunnel through it. Of the electrons which arrive at the barrier something like one part in 10^9 will tunnel through. First of all, I should like to describe how one makes these specimens in practice. One way of doing it is to use thin film techniques. One evaporates a metal such as tin or lead through a mask and obtains a superconducting film which one then exposes to the air for a few minutes so that it grows an insulating coat of oxide. One then evaporates a second strip of superconductor across it, thus obtaining a layer of metal, insulator, and then metal again. Finally one attaches electrodes and measures the current–voltage characteristic of the layer of insulator. Other techniques have also been found for making these devices; for example, you can simply press together two pieces of niobium wire, and the oxide layer which is naturally present on top of the niobium is about the right sort of thickness to make the type of device needed.

Figure 1 shows a typical example of a characteristic obtained from these devices. Plotted here is the current through the barrier as a function of the voltage across it. As you can see, the characteristic has two parts.

> The author was visiting the Department of Physics, University of Illinois, Urbana, Illinois, at the time of the Conference; his permanent address is Cavendish Laboratory, University of Cambridge, Cambridge, England.

In one part the voltage across the barrier is zero, and yet there is still current flowing through it. This, being a current without voltage, is a supercurrent just like the sort of current which flows in a bulk piece of superconducting wire. In the second portion of the characteristic there is a finite voltage and the barrier is behaving as if it is resistive. This part of the characteristic is non-linear. The theories of the two parts are completely different, and I shall spend about half the lecture talking about each.

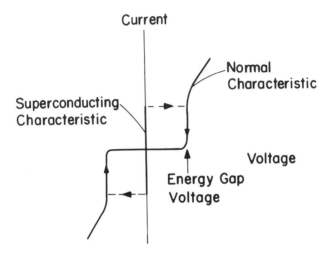

Fig. 1 Schematic current–voltage characteristic of a superconducting tunnel junction at very low temperature.

First of all, let me explain the difference between these two pieces of characteristic. As described in the first chapter, the superconducting state is a state in which there is pairing of the electrons. At absolute zero, to a first approximation, there is complete pairing, and if a particular electron state is occupied, then the state of opposite momentum and spin is also occupied. At finite temperatures some of these pairs are broken up, and we describe this in terms of quasi-particles or excitations moving about in the superconductor. A state of the system in which a supercurrent is present is one in which one has a modified form of pairing in which the momentum of each pair is no longer zero. In other words, a supercurrent is a flow of pairs, and does not involve the flow of excitations.

On the other hand, a normal current does involve the flow of excitations. This involves making a non-equilibrium situation, and there is dissipation. In the zero-voltage part of the characteristic only a supercurrent is present [2] while in the other, non-linear, part normal currents are flowing as well.

I shall now indicate how one obtains a formula giving the variation of the normal current as a function of the voltage across the barrier. To do this, one regards the current as being a process in which individual electrons go across the barrier separately, each electron taking a definite energy E with it. First we ask how much current is carried by those electrons crossing the barrier in a particular direction (say left to right) in an energy range E to $E + dE$. The tunneling process can be broken up into three steps:
  (1) An electron in the energy range E to $E + dE$ leaves the left hand piece of superconductor;
  (2) the electron tunnels across the barrier; and
  (3) the electron enters the right hand piece of superconductor.
The formula for the current

$$I_{l \to r, \, E \text{ to } E+dE} = (2\pi/\hbar)|T|^2 (dN/dE)_l (dN/dE)_r n_l(1-n_r) \, dE$$

contains three factors, one coming from each step of the tunneling process. The first factor,

$$(dN/dE)_l,$$

is the density of states on the left-hand side. This represents essentially the number of electrons available of a given energy E on the left-hand side. Then there is a factor which gives the probability of the electrons going across the barrier. According to quantum mechanics this is proportional to a squared matrix element $|T^2|$. Finally one has a factor representing the probability that the electrons will find a state of appropriate energy on the second side, giving a factor $(dN/dE)_r$, the density of states on the right-hand side. There is an additional factor which takes into account the exclusion principle. The electron can leave the left side only if there is an occupied state on that side and it can end up on the other side only if there is an unoccupied state it can go into. This gives the factor $n_l(1-n_r)$, n being the Fermi factor. To get the total current, one integrates with respect to E, giving the current in one direction, and subtracts a similar term, giving the current in the opposite direction. The final result is

$$I = (2\pi/\hbar) \int |T|^2 (dN/dE)_l (dN/dE)_r (n_l - n_r) \, dE$$

Now, the important thing about this is that it contains the densities of states. This is why measurement of current-voltage characteristics in tunneling specimens is such a useful tool. In a normal metal the density of states is almost constant in the energy range of interest.

However, in a superconductor the density of states is by no means constant; there are no states at all inside the energy gap, while outside the gap there is an increased density of states.

When the formula for the density of states is inserted into the expression for the tunneling current, various types of characteristic result. Figure 1 is an example of the characteristic for superconductors at very low temperatures. Virtually no current flows until a particular voltage is reached, and that voltage is a direct measure of the energy gap. Essentially, no current flows until enough energy is supplied by the voltage across the barrier to start breaking up pairs.

Concerning the applications of tunneling to the study of superconductivity, first of all it is a method giving you directly the value of the energy gap of a superconductor. And this is something which was very difficult to do before the technique of tunneling was available. It involved looking at absorption in the far infrared region, which is a very difficult experiment to do.

If one goes to a finite temperature, one obtains a different kind of characteristic, because quasi-particles are present and extra contributions to the current come from the tunneling of quasi-particles. What theory predicts ideally is a sharp rise in current as before, when the voltage across the barrier is equal to the sum of the energy gaps in two sides. This is the voltage required to break up a pair. In addition, when the voltage is equal to the difference in two energy gaps, in theory the current is infinite, although in practice the current is finite because the gap is not infinitely sharp. A typical experimental curve such as has been obtained by Giaever would resemble that shown in Figure 2. One feature of this characteristic, important from the technological viewpoint, is that it has a negative resistance region. In general negative resistance devices can be used both to oscillate and to amplify, and indeed a barrier of this type has been used as an oscillator. Attempts have also been made to use them as amplifiers, without much success as far as I know.

Having measured the energy gap of a superconductor, one can go on to study things like its temperature variation. Some recent studies found that there were deviations from the temperature variations predicted by the BCS theory, but people have now produced more sophisticated strong-coupling theories which take these deviations into account. The theory seems now to be confirmed to a quite high degree of accuracy. One prediction of the theory is that some features of the electron–electron interaction responsible for superconductivity should show up in the tunneling characteristics. These effects are particularly striking in the case of lead, where they give rise to kinks in the characteristic big enough to be visible at a glance on an oscilloscope trace. To explain this, let me quote a formula for the density of states which was derived by

Scalapino, Schrieffer and Wilkins [3], using the strong-coupling theory. The result was that the density of states in a superconductor is equal to the real part of $E/\sqrt{E^2-\Delta^2(E)}$. Δ is again the energy gap parameter: however, in the strong-coupling theory it is not a constant as is assumed in the BCS theory, but is a function of energy instead. Notice, incidentally, that if Δ *is* assumed to be constant, this formula reduces to the BCS formula. According to the strong-coupling theory, Δ involves the interaction between the electrons, which interaction is mediated by phonons. Consequently, singularities in the phonon spectrum reappear in the density of states. With materials like lead, considerable structure is revealed if one plots the second derivative of the characteristic electronically. The location of the structure correlates well with that deduced from the phonon density of states obtained by inelastic neutron scattering. We have here very good evidence that phonons give rise to superconductivity.

I should now like to mention other applications of tunneling using the resistive parts of characteristics before I go on to discuss supercurrents. One type of measurement, which was pioneered by Zavaritskii [4],

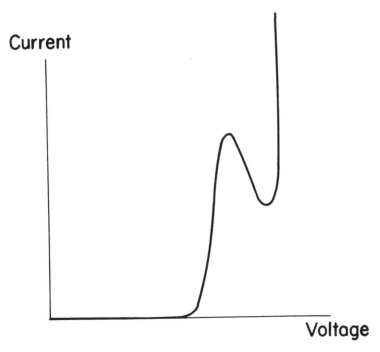

Fig. 2 Schematic current voltage characteristic of a superconducting tunnel junction at higher temperature.

is the determination of the anisotropy of the energy gap of a superconductor. Zavaritskii was able to make specimens in which the superconductor on one side was a single crystal. Because of the way in which the probability of tunneling depends on the angle of incidence of the electrons, one is selecting electrons which are traveling almost perpendicular to the face of the barrier. Hence, if there is any anisotropy, one can observe it by looking at the characteristics produced by tunneling from crystal faces of different crystallographic orientations. Zavaritskii was able not only to look at the anisotropy of the gap but also in certain directions to see more than one gap. There is a vast amount of information there, but I do not think very much has been done in the way of theoretical analysis of anisotropy to explain all these results.

Another application I should like to mention is to gapless superconductivity. It was predicted by Abrikosov and Gor'kov [5] that if magnetic impurities are added to a superconductor the transition temperature drops down quite rapidly and, with a couple of percent of impurity, the material becomes non-superconducting. Abrikosov and Gor'kov also predicted that the energy gap would go to zero before the material became non-superconducting. This can be tested by looking at the tunneling characteristics, and Reif and Woolf [6] found that one can indeed have a characteristic in which there is no energy gap even though the electrical resistance of the film is still zero.

I shall now go on to the theory of supercurrents through barriers, which give rise to the part of the characteristic at zero voltage. Let me remind you again that a supercurrent flows when the pairing is modified and the correlated pairs of electrons no longer have zero net momentum. That there is a supercurrent flowing is related to the fact that the phase in the superconductor is varying as a function of position. In fact if the momentum of the pair is $\hbar\mathbf{k}$, then the wave function is $\psi = \exp(i\mathbf{k}\cdot\mathbf{r})$. So the momentum of a pair, and hence also the value of the supercurrent, is proportional to the gradient of the phase of ψ, which can therefore be considered to be a driving potential for the superconducting electrons. Now a barrier with superconductor on both sides can really be regarded as a very dirty region inside a single piece of bulk superconductor, and as long as there is not too much of it (i.e., if it is thin enough) it is still possible for supercurrents to flow through it. Just as the direction of current flow in a bulk superconductor is determined by the way the phase of the superconducting wave function varies from place to place, the supercurrent through a barrier is determined by how much the phase of the wave function differs from one side of the barrier to the other. This phase difference is usually denoted by ϕ, and the total current through the barrier turns out to be proportional to its sine: $I = I_1 \sin \phi$.

If the phase of the wave function is different in different parts of the barrier, this formula is no longer applicable, and one has instead a formula for the current density:

$$j = j_1 \sin \phi \tag{1}$$

Since the maximum possible value of $\sin \phi$ is 1, j_1 is simply the critical current density.

Now one can go on to ask what happens if one applies electric or magnetic fields to the barrier. The answer is contained in equation (1) in conjunction with the following set of equations:

$$\left. \begin{aligned} \frac{\partial \phi}{\partial x} &= \frac{2ed}{\hbar c} H_y \\ \frac{\partial \phi}{\partial y} &= -\frac{2ed}{\hbar c} H_x \end{aligned} \right\} \tag{2}$$

$$\frac{\partial \phi}{\partial t} = \frac{2e}{\hbar} V \tag{3}$$

$$\frac{\partial H_y}{\partial x} - \frac{\partial H_x}{\partial y} = \frac{4\pi j}{c} + \frac{4\pi C}{c} \frac{\partial V}{\partial t} \tag{4}$$

In these equations the barrier is assumed to occupy the xy-plane; d is the thickness of the magnetic field region near the barrier; V is the voltage across the barrier; and C is the capacitance of the barrier per unit area. Equations (2) state that the effect of a magnetic field $\mathbf{H}$ is to cause the phase to vary in position. Equation (3) states that the effect of applying a voltage across the barrier is to make the phase depend on time. Finally, equation (4) is essentially Maxwell's equation, curl $\mathbf{H} = 4\pi \mathbf{j}/c + \partial \mathbf{D}/c\, \partial t$. Equations (1–4) form a complete set of equations for describing the barrier (ignoring resistive effects altogether). To make things simpler, one can eliminate everything except the phase, and one gets the equation

$$\nabla^2 \phi - (1/v^2)(\partial^2 \phi/\partial t^2) = \lambda^{-2} \sin \phi \tag{5}$$

where v equals $c/(4\pi dC)^{1/2}$, which is of the order of 2×10^9 cm/sec and λ is $(\hbar c^2/8\pi j_1 ed)^{1/2}$ which is of the order of $\frac{1}{2}$ mm.

Now let me return to equations (2) and (3) and discuss their origin briefly. Normally the wavelength of an electron is equal to h/mv, where v is its velocity. In a magnetic field, however, this has to be changed to $h/(mv + e\mathbf{A}/c)$, where $\mathbf{A}$ is the vector potential. Thus a magnetic field has an effect on the wavelength of an electron, and hence on the way its phase varies from place to place. This statement so far applies only to electrons in free space. However, according to the Ginzburg–Landau theory of superconductivity a similar effect occurs in superconductors, and this theory leads one to equation (2). Note that equation (2) contains the combination 2*e*, which is a result of the fact that superconductivity involves pairs of electrons rather than isolated electrons.

Equation (3) is a consequence of the quantum-mechanical energy-frequency relation $E = \hbar\omega$. If a potential difference V is applied across a barrier, electron pairs have an energy differing by $2eV$ on the two sides of the barrier. This corresponds to a frequency difference $\omega = 2eV/\hbar$, and this is just the rate of change of phase $\partial\phi/\partial t$ which appears in equation (3).

Now, I want to describe some experimental observations supporting the theory I have described. The first is an experiment that was carried out by Anderson and Rowell [7] at Bell Labs. What they did was to apply a magnetic field to the barrier, and measure the critical current of the barrier as a function of magnetic field. They found an oscillatory variation of critical current, which can be explained as follows. If we do not apply a magnetic field to the barrier, the phase is independent of position (eq. 2). The current, which depends on the sine of the phase, is also independent of position, so the current distribution will be as in the top diagram (Fig. 3). If we apply a magnetic field to the barrier,

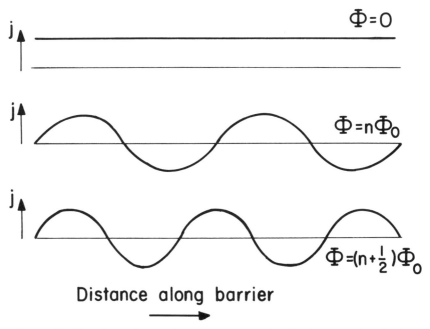

Fig. 3 Position dependence of the supercurrent density *j* across a barrier for three values of the flux Φ contained within the barrier. Φ_0 is the unit of flux, and *n* is an integer.

it will make the phase vary linearly with position. The current will therefore oscillate sinusoidally (bottom two diagrams). The total current through the barrier is clearly less in the presence of a magnetic field, and when the applied field is a multiple of a particular field H_0, the critical current of the barrier is actually zero (bottom diagram). The critical current is in general proportional to $|\sin \pi(\Phi/\Phi_0)|/(\Phi/\Phi_0)$, where Φ and Φ_0 are the flux contained within the barrier and the quantum of flux respectively.

The mathematics involved here is very similar to that of the theory of the diffraction grating, and the experiment can be regarded as an interference experiment just like those that can be carried out with isolated electrons or with light. One can also do an analogy of a two-slit diffraction experiment by using two barriers connected in parallel instead of a single barrier. The superconducting electrons going through the two barriers interfere with each other, with the result that the critical current of the combination oscillates as the amount of flux enclosed between the two barriers is varied. This effect can be used to measure very small magnetic fields [8].

The experiments I have discussed show the effects of the magnetic field in causing the phase to vary over the barrier. Now I shall describe some experiments which show the effects of a potential difference, which causes the phase to vary with time. First of all, if one simply applies a steady potential difference across the barrier, that will make the phase vary linearly with time. The current, which is equal to j_1 times the sine of the phase, will therefore oscillate with time, the frequency being given by $\hbar\omega = 2eV$. This is the ac supercurrent. A barrier is the simplest form of oscillator yet invented; if you simply apply a dc voltage to a barrier, currents will oscillate back and forth across it. Numerically the voltage–frequency relation implies that if a voltage of one millivolt is applied to a barrier, the ac supercurrent oscillates at a frequency of 500 gigacycles, and the radiation produced is coherent. A technical problem which has not been solved yet is getting the radiation out efficiently. So far the maximum power that has been extracted is 10^{-11} watts, though it is hoped that this figure can be increased by using different types of junctions.

The ac supercurrent under some circumstances gives rise to interesting effects on the characteristic of a barrier. This happens when the barrier behaves like a resonant cavity. As the voltage across the barrier is altered, the frequency of the ac supercurrent changes, and every time it becomes equal to one of the resonant frequencies of the barrier a step appears in the current–voltage characteristic. The interpretation of the steps is as follows. When the frequency of the ac supercurrent reaches that of a resonance, the amount of power dissipated inside the barrier increases. This power can only be supplied by the dc source, so that the amount of current through the barrier increases.

A somewhat more interesting experiment consists of placing the barrier inside a cavity containing externally generated microwave radiation. It is possible under certain circumstances to lock the frequency of microwaves to the frequency of the ac supercurrent. This can happen whenever the ratio between the two frequencies is a simple rational number. It is found that whenever frequency-locking occurs a step appears in the characteristic of the barrier. By measuring simultaneously the voltage of the step and the frequency of the microwaves, a very accurate measurement of the fundamental constant h/e which enters into the voltage–frequency relation can be obtained. An experiment to do this is now being carried out at the University of Pennsylvania. [Editor's note: See Parker, W. H., Taylor, B. N., and Langenberg, D. N., 1967. *Phys. Rev. Letters* 18: 289.]

I should like finally to mention an effect which has not yet been observed, but which is interesting because it is the analogue of the plasma oscillations which occur in a bulk metal. Consider again equation (5):

$$\nabla^2 \phi - (1/v^2)(\partial^2 \phi / \partial t^2) = \lambda^{-2} \sin \phi$$

Let us suppose that ϕ is independent of position and small, so that $\sin \phi$ can be approximated by ϕ. The equation then becomes:

$$d^2\phi/dt^2 + \omega^2 \phi = 0$$

where $\omega = v/\lambda$. It therefore describes oscillation at frequency ω, which is of the order of a few gigacycles. It may be considered as a plasma oscillation, whose frequency is very low because the density of charge carriers in the barrier is very small.

To summarize then, tunneling is a very useful tool. If one looks at the resistive characteristic, one can find out many kinds of information such as energy gaps, densities of states, interactions responsible for superconductivity, etc. By looking at the supercurrent part, on the other hand, one learns about the coherent properties of the superconducting state and one can observe quantum-mechanical effects on a macroscopic scale.

References

[1] Giaever, I. 1960. *Phys. Rev. Letters* 5: 147.
[2] Josephson, B. D. 1965. *Advances in Phys.* 14: 419.
[3] Scalapino, D. J., Schrieffer, J. R., and Wilkins, J. W. 1963. *Phys. Rev. Letters* 10: 336.

[4] Zavaritskii, N. V. 1962. *Zhur. Eksp. Teor. Fiz. SSSR* 43: 1123 (translation: *Soviet Phys. JETP* 16: 793).
[5] Abrikosov, A. A., and Gor'kov, L. P. 1960. *Zhur. Eksp. Teor. Fiz. USSR* 39: 1781 (translation: *Soviet Phys. JETP* 12: 1243).
[6] Reif, F., and Woolf, M. A. 1962. *Phys. Rev. Letters* 9: 315.
[7] Anderson, P. W., and Rowell, J. M. 1963. *Phys. Rev. Letters* 10: 230.
[8] Mercereau, J. E. 1967. Present volume, p. 63.

3
Superconducting Materials
Ted G. Berlincourt

Introduction

In this paper I shall first define some general classifications of bulk, homogeneous superconducting materials. It will be convenient to discuss these classifications within the framework of the Ginzburg–Landau (GL) theory and its extensions [1–5]. I shall then describe some of the profound effects of geometrical modifications on the magnetic characteristics of homogeneous materials. Heterogeneous structures and inhomogeneities are next considered. A discussion of the factors which govern the occurrence of superconductivity follows, and examples of a number of typical and atypical materials are cited. Finally, I shall note some of the more speculative theoretical advice on where to look for superconductivity.

The Ginzburg–Landau Theory and Its Extensions

In discussing the GL theory as a basis for categorizing superconductors I shall refer frequently to the "degree of superconducting order" or the "superconducting order parameter" Ψ. This quantity may be interpreted as a measure of the energy gap

$$\Delta \propto \Psi, \qquad (1)$$

as a measure of the superelectron pair density

$$n_p \propto |\Psi|^2, \qquad (2)$$

The author is associate director of the Science Center of the North American Rockwell Corporation, Thousand Oaks, California.

or simply as a measure of the "strength" of superconductivity at a particular point. In a normal metal Ψ is of course zero. The GL theory pictures a superconductor as a flexible physical system—one which responds to applied currents and magnetic fields by adjusting its spatial distribution of order. The resulting configuration of order, current, and field is then the one which minimizes the total energy of the system. We shall see that there exist at least three general types of configurations or "phases" depending upon applied magnetic field strength and certain material parameters.

According to the GL theory, near the superconducting transition temperature T_c the energy difference $f_s(H_a) - f_n$ between a superconducting configuration and a normal one in an applied magnetic field H_a is given by terms related to the magnitude of the order, plus a magnetic field energy, plus the kinetic energy of the electrons, including the energy of interaction of the current with the field; i.e.,

$$f_s(H_a) - f_n = \underbrace{\alpha|\Psi|^2 + \frac{1}{2}\beta|\Psi|^4}_{\text{Order}} + \underbrace{\frac{H^2}{8\pi}}_{\text{Field}} + \underbrace{\frac{1}{2m^*}\left| -i\hbar\nabla\Psi - \frac{e^*}{c}A\Psi \right|^2}_{\text{K.E., JH Interaction}}. \quad (3)$$

Here α and β are temperature dependent material parameters, e^* and m^* are twice the electronic charge and mass respectively, and $\hbar$ and c have their usual meanings. The quantum mechanical momentum operator for a charged particle in a magnetic field will be recognized in the last term. For constant temperature T only two variables appear in this equation, Ψ and the field (or the vector potential $\mathbf{A}$, which is related to the magnetic field by the expression $\mathbf{H} = \nabla \times \mathbf{A}$). To find the equilibrium conditions, Ginzburg and Landau minimized the condensed-phase energy f_s with respect to Ψ and also with respect to $\mathbf{A}$. This yielded the GL equations

$$\frac{\partial f_s}{\partial \Psi^*} = 0: \quad \left(\frac{i}{\varkappa}\nabla + \mathbf{A}\right)^2 \Psi - \Psi + \Psi|\Psi|^2 = 0, \quad (4)$$

$$\frac{\partial f_s}{\partial \mathbf{A}} = 0: \quad \nabla^2 \mathbf{A} = \frac{1}{2\varkappa}(\Psi^*\nabla\Psi - \Psi\nabla\Psi^*) + |\Psi|^2 \mathbf{A}. \quad (5)$$

As written here in the usual GL *reduced units*, these equations involve only Ψ, $\mathbf{A}$, and an important parameter $\varkappa$, which is a constant characteristic of a particular material. The appropriate boundary condition

$$\mathbf{v} \cdot \left(-i\hbar\nabla\Psi - \frac{e^*}{c}\mathbf{A}\Psi \right) = 0 \quad (6)$$

simply requires that the normal component of electron momentum vanish at a metal–vacuum interface. ($\mathbf{v}$ is a unit vector normal to the interface.)

In general we are interested in spatial distributions of Ψ and $\mathbf{A}$ which satisfy the two coupled equations (4) and (5) and the boundary condition (6). For different values of $\varkappa$ there exist different classes of solutions for Ψ and $\mathbf{A}$, and so it is important to have a feeling for the influence of $\varkappa$ and also for the material parameters that determine $\varkappa$. The quantity $\varkappa$ is a measure of the spatial "flexibility" of the order in a specimen in the presence of an applied magnetic field. For $\varkappa = 0$ the order is constant in space as in the Londons' theory.[1] For small $\varkappa$ the order is relatively "stiff" or "inflexible." On the other hand, for large $\varkappa$ the order is very flexible and is readily deformed by a magnetic field. GL expressed the material parameter $\varkappa$ as

$$\varkappa = \frac{\sqrt{2e^*}}{\hbar c} H_c \lambda_0^2, \quad (7)$$

where λ_0 is the low-magnetic-field penetration depth, and H_c is the "thermodynamic critical field." ($H_c^2/8\pi$ is the zero-field condensation energy per unit volume for a superconductor.) The GL equations have been derived from the Bardeen–Cooper–Schrieffer (BCS) microscopic theory [6] in the local limit by Gor'kov [3, 4]. His expression for $\varkappa$ in terms of measurable parameters of the *normal* state may be approximated by the relation

$$\varkappa \approx 1.61 \times 10^{24} \frac{T_c \gamma^{3/2}}{n^{4/3}(S/S_f)^2} + 7530 \rho_n \gamma^{1/2}, \quad (8)$$

where T_c is the superconducting transition temperature, γ is the normal-state electronic specific heat coefficient (ergs/cm^3 deg^2), n is the normal-state valence electron density (cm^{-3}), S/S_f is the ratio of the Fermi surface area to that for a free electron gas of density n, and ρ_n is the normal-state electrical resistivity (ohm cm). From this relation it is evident that a very effective way to increase $\varkappa$ is to add impurities which shorten the electron mean free path and thus increase ρ_n.

Bulk Homogeneous Superconductors

We now consider the various types of solutions of the GL equations which follow from different values of the material parameter $\varkappa$. We discuss the case of a thick plate of infinite extent in two dimensions, with an applied magnetic field parallel to the surface of the plate. Three ranges of $\varkappa$ are of interest because they correspond to different types of superconductors. The energies given by the GL equations for these three

---

[1] It should be noted, however, that in contrast with the Londons' theory, the GL theory yields a magnetic-field-dependent order.

types of superconductors are illustrated very schematically as functions of magnetic field strength in Figure 1. For small $\varkappa$ ($\varkappa < 1/1.69\sqrt{2}$) there

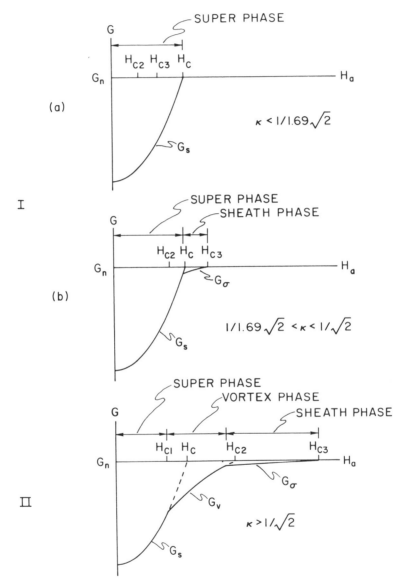

Fig. 1 Dependence of energy on applied magnetic field shown schematically for the three types of superconductors. The energy branches corresponding to the "super," "sheath," and "vortex" phases are designated respectively by G_S, G_σ, and G_V.

exists only a single condensed-phase energy branch G_s corresponding to a nearly perfectly diamagnetic or "super" phase. This super phase transforms to the normal state at the thermodynamic critical field H_c. We shall identify such a material as a Type I(a) superconductor.

For intermediate values of $\varkappa$ ($1/1.69\sqrt{2} < \varkappa < 1/\sqrt{2}$) a second condensed-phase energy branch G_σ appears above H_c. It corresponds to the "sheath" phase of Saint James and De Gennes [5] and extends to a field H_{c3} where the normal state is restored. We shall identify such a material as a Type I(b) superconductor.

For large $\varkappa$ ($\varkappa > 1/\sqrt{2}$) a third energy branch G_V appears between the super and sheath branches. It corresponds to the "mixed" phase or "vortex" phase of Abrikosov [2] and extends between a "lower critical field" H_{c1} and an "upper critical field" H_{c2}. We identify such a material as a Type II superconductor.

The realms of existence of the different phases are indicated in another way in Figure 2. On the left, magnetic field is plotted against $\varkappa$ for

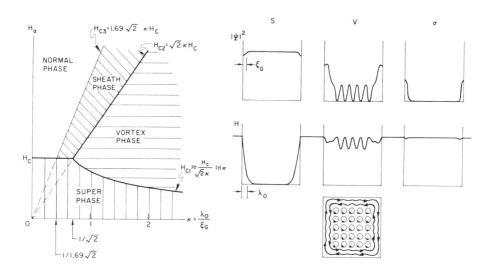

Fig. 2 On the left magnetic field is plotted against $\varkappa$ for constant condensation energy. The boundaries between phases correspond to the various critical fields. On the right the distributions of order and field for the various phases are indicated very schematically.

constant condensation energy, and the boundaries between phases correspond to the various critical fields. On the right the distributions of order and field for the various phases are indicated very schematically.

Consider first a large $\varkappa$ or Type II material ($\varkappa > 1/\sqrt{2}$). At low fields the nearly perfectly diamagnetic super phase exists. Within a characteristic distance, the penetration depth λ_0, the magnetic flux density drops to $1/e$ of its value outside the specimen and then rapidly approaches zero. At the same time the order is degraded in a region of thickness ξ_G near the surface as indicated. The characteristic distance ξ_G is known as the "coherence length." Together with λ_0 it provides a rough estimate of $\varkappa$ as

$$\varkappa \approx \frac{\lambda_0}{\xi_G}. \tag{9}$$

As the field applied to the super phase of the Type II superconductor is increased, the increasing magnetic pressure associated with perfect diamagnetism becomes very uncomfortable. When the field H_{c1} is reached, the Type II superconductor, which has a very flexible order, is willing to sacrifice some of that order in exchange for a relief of magnetic pressure. In so doing it enters the vortex phase, which is characterized by the periodic spatial distributions of order and field shown schematically under V in Figure 2. Flux entry is associated with little supercurrent whirlpools or single-quantum supercurrent vortices as indicated. At the center of each vortex the order vanishes, and the field reaches its maximum value. The mutual magnetic interactions between vortices lead to a lattice-like array or crystallization. Neutron scattering experiments have revealed a triangular array for the vortex state in Nb [7].

As the field applied to a Type II superconductor is increased further, the mean level of the order in the vortex phase is gradually depressed, until at the field H_{c2} the order vanishes entirely well inside the specimen. However, there remains a solution to the GL equations corresponding to the sheath phase, which has the order and field distributions shown schematically at the far right in Figure 2. As indicated, the presence of a metal-vacuum interface leads to the survival of superconducting order in a layer of thickness $\sim \xi_G$ at the surface. Although this layer is too thin to distort the applied field very appreciably it is easily detected by tunneling experiments [8, 9] and surface impedance measurements [10], both of which look only at the surface. Finally, for fields greater than H_{c3} the normal state appears throughout.

For intermediate values of $\varkappa$ ($1/1.69\sqrt{2} < \varkappa < 1/\sqrt{2}$), the order is so rigid as to preclude appearance of the vortex phase, and, as the field is increased, a transition takes place directly from the super phase to the sheath phase, followed by the normal phase. Pure Pb is an example of such a Type I(b) material.

For the smallest values of $\varkappa$ ($\varkappa < 1/1.69\sqrt{2}$), only the super phase appears. Nevertheless the field H_{c3} retains significance. The equilibrium transition at H_c is a first order transition with a latent heat, and therefore a phenomenon analogous to the supercooling of a liquid below its normal freezing point may take place. In the present instance, as the field applied to a Type I(a) superconductor is decreased through H_c, the normal state may persist in a metastable condition. However, when the field is lowered to H_{c3} nucleation of superconductivity *must* commence at the surface, and we then expect superconducting order to spread through the specimen. Thus in a Type I(a) superconductor H_{c3} is the "supercooling" limit of the normal phase.

The theoretical dependences on $\varkappa$ of the transition fields shown in the phase diagram of Figure 2 are as follows:

$$H_{c1} \approx \frac{H_c}{\sqrt{2\varkappa}} \ln \varkappa, \qquad (\varkappa \gg 1), \tag{10}$$

$$H_{c2} = \sqrt{2\varkappa} H_c, \tag{11}$$

$$H_{c3} = 1.69\sqrt{2\varkappa} H_c = 1.69 H_{c2}. \tag{12}$$

H_c is of course independent of $\varkappa$ in Figure 2 in keeping with the intitial assumption of constant condensation energy.

In Figure 3 are shown very schematically the dependences of average

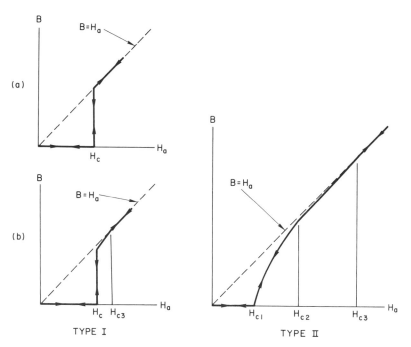

Fig. 3 Average internal field B versus applied magnetic field H_a shown very schematically for the three types of superconductors. Departures of the sheath phase branches from the $B = H_a$ lines have been exaggerated.

internal magnetic field B on the applied magnetic field H_a for the three categories of superconductors we have considered. Perfect reversibility has been assumed, and the normal-state branches are shown following the $B = H_a$ lines corresponding to the assumption of negligible normal state magnetism. The super phase in all three cases is indicated by the perfectly diamagnetic $B = 0$ branches. The gradual penetration of flux in the mixed state is shown for the Type II superconductor. For the sheath phase branches the departures from the $B = H_a$ line have been grossly exaggerated.

As mentioned earlier, the results we have been discussing are subject to the restriction of the GL theory that the temperature be very close to the transition temperature. When this condition is not satisfied the situation becomes more complicated, as indicated both by experiment and the microscopic theory [11–17]. Not only is $\varkappa$ temperature-dependent, but in addition, if we wish to retain the expressions (10) and (11)

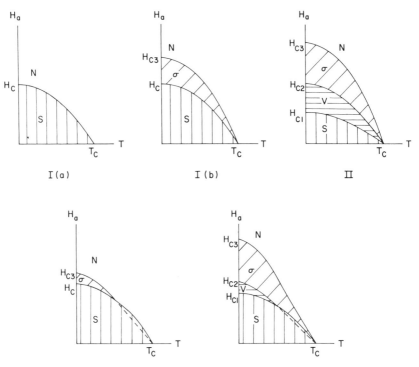

Fig. 4 Schematic representations of the realms of existence of the various phases of superconductivity in the magnetic field–temperature plane for the various types of superconductors. The lower diagrams result when $\varkappa$ changes with temperature in such a way as to change the type designation.

presented above for H_{c1} and H_{c2}, it is necessary to define separate $\varkappa$ values for H_{c1} and H_{c2}. This leads to the interesting situation illustrated in Figure 4, where the realms of existence of the various phases in the field-temperature plane are shown very schematically. The phase diagrams for simple Type I(a), I(b), and II superconductors appear at the top. At the lower left is shown the phase diagram for a material which changes from Type I(a) to Type I(b) as $\varkappa$ increases with decreasing temperature. The phase diagram at the lower right illustrates a transition from Type I(b) to Type II as $\varkappa$ increases with decreasing temperature.

We now consider another important departure from the elementary GL theory. Many materials possess $\varkappa$ values which are so large that their upper critical fields are of the order of 100 kG. Moreover, the normal states of these large $\varkappa$ materials are often quite strongly paramagnetic. In such instances the magnetic energy of the normal state $-\frac{1}{2}\chi_N H_a^2$ (where χ_N is the normal state magnetic susceptibility) must be taken into account. In addition, the spin paramagnetism of the unpaired electrons in the condensed phase must be considered. The resultant effect on the energy branches is illustrated very schematically at the top of Figure 5. In contrast to the simple case shown in Figure 1 the normal-state energy G_N decreases parabolically with applied field, and the vortex-state energy G_V goes through a maximum before meeting G_N. (For the present purposes we ignore the sheath phase.) The corresponding curves of average internal field versus applied field are shown schematically at the bottom of Figure 5. The region above the $B = H_a$ line is paramagnetic, and the region below the $B = H_a$ line is diamagnetic. We note that with increasing field the vortex state changes from diamagnetic to paramagnetic as the vortex-state energy curve goes through its maximum. Stated in somewhat superficial terms this vortex state paramagnetism sets in when the spin paramagnetism of the unpaired electrons exceeds the diamagnetism of the supercurrents. Direct magnetic and caloric evidence for such behavior has recently been obtained [18–20], and the appropriate microscopic theory has also been developed [14, 17].

We now turn to a consideration of the responses of the various condensed phases of bulk homogeneous superconductors when dc electric transport currents are imposed upon them in the presence of applied magnetic fields. In the super phase, current flow is confined to the penetration depth. This current flows without dissipation until the vector sum of the applied field and the field due to the transport current equals the transition field which marks the limit of the super phase (H_c for Types I(a) and I(b), and H_{c1} for Type II). Critical transport current densities within the penetration layer are quite large, typically of the order of 10^7 to 10^8 A/cm^2 for zero applied field.

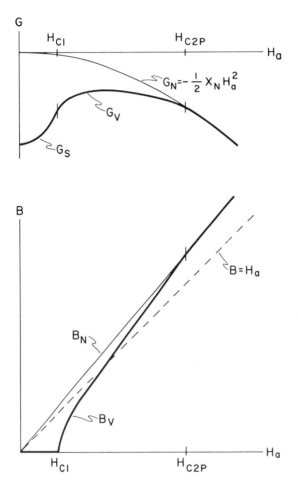

Fig. 5 At the top are shown very schematically the field dependences of the energy branches G_S, G_V, and G_N for the super, vortex, and normal phases of a superconductor characterized by a non-negligible normal-state magnetism. The average internal magnetic flux densities corresponding to these energy branches are shown as functions of the applied field in the lower figure.

In contrast, the pure vortex state is unstable against perturbations. The situation is illustrated in Figure 6 where a current-carrying rod is shown in a transverse field sufficient to induce the vortex state. For zero transport current the flux lines and supercurrent vortices thread the cross section as shown at the lower left. With finite transport current, the field due to the current distorts the flux distribution as shown schematically at the lower right in Figure 6. Now in a perfectly homogeneous material there is no reason for a vortex to prefer any one location over any other. Hence the vortices are free to move to the right in an effort to homogenize the flux distribution. Vortices are formed on the left side, and they migrate to the right side where they are annihilated. This so-called "flux flow" is a dissipative process [21]; i.e., electric fields generated by this motion cause current to flow through the cores

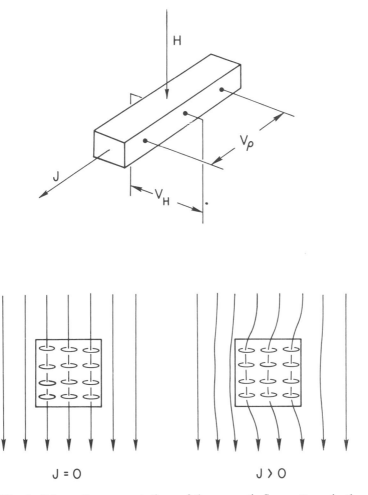

Fig. 6 Schematic representations of the magnetic flux patterns in the vicinity of a vortex state bar in a transverse applied field with and without an applied transport current.

of the vortices, which are in effect normal [22]. A voltage V_ρ is thus developed along the length of the specimen. In addition, there is some evidence that vortex motion is not precisely transverse to the specimen, thereby giving rise to a transverse or Hall voltage V_H [23–26]. As each vortex traverses the specimen it carries an effectively normal core and can therefore be thought of as transporting entropy across the specimen. As a consequence thermoelectric effects analogous to the Peltier, Ettingshausen, and Nernst–Ettingshausen effects are observed under flux flow conditions. In the Peltier-type experiment heat is either emitted or absorbed at a super phase–vortex phase junction depending on the direction of current flow [27]. In the Ettingshausen-type experiment a transverse temperature gradient is produced by the current, and in the Nernst–Ettingshausen-type experiment a transverse voltage is produced by a thermal gradient [28]. The latter two effects are about three orders of magnitude larger than those observed for normal metals.

Because we have thus far been concerned only with perfectly homogeneous material, we might expect flux flow to commence in the vortex state for vanishingly small currents, but in fact appreciable currents are required even for the most homogeneous specimens. This is a consequence of the special character of a metal–vacuum interface. We know of course that the surface is very special; i.e., the strength of superconductivity is greatest at the surface, for otherwise there would be no sheath phase. Accordingly, in the vortex phase as the surface is approached from the inside, the periodic structure of the interior is matched to a strong superconducting sheath [29, 30]. If flux is to enter through this sheath a small normal region must first puncture the sheath to serve as the vortex core. A certain threshold force is required for this, and so a finite transport current is required to initiate flux flow even in a presumably perfectly homogeneous specimen [31]. The vortex state sheath also gives rise to important effects when no applied transport current is present; e.g., if the vortex state sheath is multiply connected (as in the cross section shown in Figure 2) it inhibits attainment of the equilibrium magnetization state [32–34]. Irreversibilities also occur in the pure sheath state (between H_{c2} and H_{c3}) whenever the geometry is such that the sheath is multiply connected [35]. Several theoretical and experimental investigations of the sheath critical current have been carried out [36–40].

Geometrical Effects in Homogeneous Material

In this section we consider how geometry affects the various phases of superconductivity and how it may produce extreme changes in the character of superconducting materials. As we have seen, bulk materials rapidly expend their condensation energy as they attempt to shield out an applied magnetic field. In contrast, specimens having dimensions comparable to the penetration depth readily admit flux to their interiors. As a consequence, condensation energy is less rapidly consumed in thin specimens, and magnetic phase transitions are moved to higher fields, or may even be suppressed entirely. In Figure 7 (which is tentative and speculative in several respects), the various transition fields of a plate in a longitudinal field are plotted very schematically against the

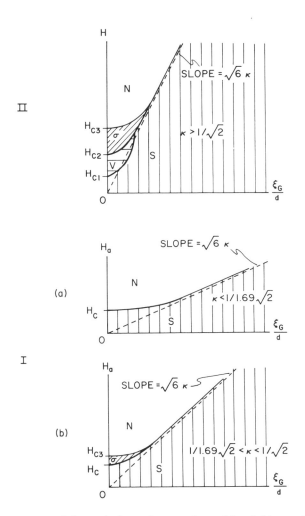

Fig. 7 Schematic dependences of transition fields on plate half-thickness d for plates of the three types of superconductors in longitudinal fields. Several features are speculative because both theory and experiment are incomplete.

coherence length divided by the plate half thickness d. In general, for all three types of superconductor, we expect the transition fields to increase as the plate thickness is decreased. Also, we expect the vortex phase of a Type II superconductor to vanish when the thickness becomes too small to accommodate a vortex. For *very* thin specimens all superconductors look alike. Because the thickness is much less than the coherence length, the order is uniform throughout the specimen. Consequently there exists only one condensed phase, the super phase, which transforms directly to the normal phase at a field which increases very rapidly with decreasing thickness. Very thin films in longitudinal fields remain superconducting in fields which are orders of magnitude greater than the transition fields of bulk specimens. In such cases normal-state magnetic energy and unpaired-electron magnetic energy must be considered as in the case of large-$\varkappa$ bulk materials. The critical transport current of a very thin film is limited by the kinetic energy of the superelectrons long before the magnetic field generated by the current becomes comparable to the zero-current transition field.

Among all the curves shown in Figure 7, the one least explored both theoretically and experimentally is that for H_{c1}, and so its speculative nature should be emphasized. Actually, the H_{c1} curve corresponds approximately to the formation of the first layer of vortices along the center plane of the plate. Additional contours could be plotted in the vortex-state region corresponding to the formation of the second, third, fourth, etc. layers. Interesting theoretical and experimental studies have been conducted on cylinders so small in diameter that only a few vortices could be accommodated within the cross section [41, 42]. The entry and exit of individual vortices were successfully detected.

Thus far we have discussed only those geometries characterized by a zero demagnetizing coefficient, geometries like infinitely long plates and cylinders in longitudinal fields. The picture presented in Figure 7 is modified considerably for nonzero demagnetizing coefficients. For example, the sheath critical field H_{c3} decreases rapidly as the applied field is given a component perpendicular to the specimen surface, and when the field is set normal to the surface the sheath vanishes completely [5, 43].

Another important configuration is the so-called "intermediate state." It occurs in bulk Type I materials having nonzero demagnetizing coefficients. (For an elementary discussion, see [44].) Consider a Type I cylinder in a transverse magnetic field as shown in cross section in Figure 8. The near-perfect diamagnetism of the super phase distorts the flux in such a way that the maximum field at the surface of the cylinder is actually twice the applied field. When this maximum field reaches H_c the specimen breaks into alternating domains of normal and super material. The scale of this structure is ordinarily large in comparison to that of the vortex state. It is determined by a compromise between the negative energy gained through relaxation of the field distortion and the positive energy paid in establishing super-normal boundaries. It is in fact the positive value of the interphase surface energy which distinguishes the Type I superconductor from the Type II superconductor and precludes establishment of the fine-scale vortex-state structure in *bulk* Type I

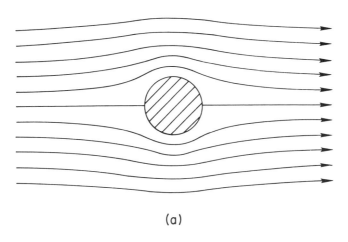

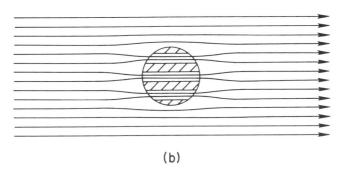

Fig. 8 (*a*) Magnetic flux distribution around the cross section of a super-state cylinder in a transverse applied field. (*b*) Intermediate-state structure showing flux penetration through normal domains.

materials. The word "bulk" is emphasized in this connection because the vortex state *is* obtained for a sufficiently thin Type I film in a *transverse* magnetic field [45–46].

Any discussion of geometrical effects should include mention of predictions for specimens having at least one dimension smaller than about 50 Å. For example, in a very thin film the quantum requirement that an integral number of de Broglie wavelengths fit within the thickness alters the electron momentum distribution markedly. The pertinent theory predicts shape resonances in the energy gap and a strong dependence of transition temperature on thickness [47–52].

Proximity Effects in Heterogeneous Structures

We now consider a different kind of geometric modification, one in which films of different materials are placed in intimate contact with one another. Certainly if we are willing to believe that a Josephson junction, two superconductors separated by a thin "insulator," can support supercurrents [53–55], then we should readily accept the fact that a thin "normal" metal layer between two superconductors can also support supercurrents. This ability of superconductors to "infect" normal metals with superconductivity is commonly called the "proximity" effect. (For pertinent references see Hauser, Theuerer, and Werthamer [56] and also Minnigerode [57].) The proximity effect is especially important because it makes available a whole new class of superconducting materials, and also because proximity effects are operative in many of the technologically important high-field materials. Several situations should be distinguished. A very thin normal metal deposited on a massive superconductor becomes superconducting. Similarly, a very thin superconductor deposited on a massive normal metal becomes normal. A composite of two thin superconducting metals exhibits a transition temperature between those of the separate components. A composite of a thin superconductor and a thin normal metal will either be normal or have a transition temperature below that of the superconductor. Parameters which determine the transition temperature of a composite are the thicknesses, transition temperatures, normal state electrical resistivities, and electronic densities of states of the components. The electron mean free path, which appears in the normal state resistivity, is especially important in composites because it controls the leakage of superconducting pairs to the unfavorable region; i.e., it controls the action depth. In some instances agreement between experiment and theory is remarkably good. Such a case is illustrated in Figure 9, where the transition temperature of a Pb–Al composite is plotted against Pb thickness for a constant Al thickness of 4400 Å [58]. Data are represented

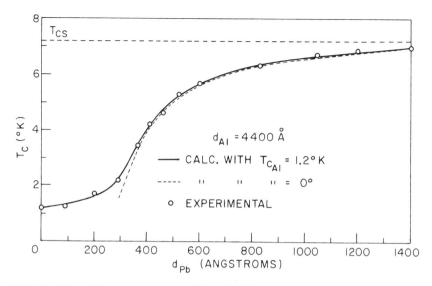

Fig. 9 Transition temperature of a Pb–Al composite versus Pb thickness for a constant Al thickness of 4400 Å [58].

by the points, and the solid line represents the appropriate theory. Ferromagnetic and antiferromagnetic overlays are particularly effective in depressing T_c. Tunneling studies have shown that a layer of Mn as little as 15 Å thick causes marked changes in the tunneling density of states as detected on the opposite side of Pb films a few thousand Å thick [59]. Normal and ferromagnetic overlays have also been used with great success to quench, or at least to degrade, the sheath both in the vortex phase and in the sheath phase [60, 31, 39]. In fact, the reduction of magnetic irreversibility in the vortex state by the application of overlay techniques has provided convincing evidence for the reality of the surface barrier in the vortex state.

Inhomogeneous Materials

We are now in a position to understand the characteristics of inhomogeneous materials of the type used in high-field supermagnets. As already mentioned, the interaction of a transport current with a transverse magnetic field leads to a force (the Lorentz force) which causes a dissipative flux-flow condition in a homogeneous Type II superconductor in the vortex state. Fortunately this dissipative process can be inhibited by the introduction of suitable inhomogeneities. These inhomogeneities attract and pin vortices, thus stabilizing the vortex state structure [61].

One type of pinning process may be understood with reference to Figure 10. At the upper left, magnetic moment is plotted against applied

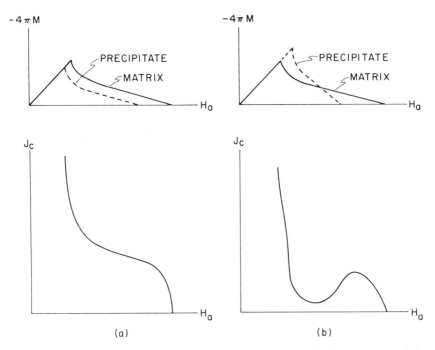

Fig. 10 (a) Schematic dependence of critical current density J_c on applied magnetic field for a mixture of materials which separately exhibit magnetization curves as shown at the top. (b) "Peak effect" in $J_c(H_a)$ arising from mixture of materials which separately exhibit magnetization curves as shown at the top.

field for two separate materials of slightly different characteristics. On such a plot the larger the quantity $-4\pi M$, the more diamagnetic the material, and the more it wants to reject magnetic flux. Suppose a composite is formed with the material represented by the solid curve as a matrix and the material represented by the dashed line as a finely dispersed precipitate. For applied fields greater than the lower critical field of the precipitate, the matrix is more diamagnetic and hence dislikes vortices more intensely than does the precipitate. As a consequence vortices are attracted to and are pinned by the precipitate particles. This situation applies even when (as at high fields) the precipitate particles are normal. The average transport current density which such a system can support before dissipation sets in is plotted very schematically at the lower left of Figure 10. At low fields the super-phase surface currents account for the high average critical current density. At higher fields the magnitude of the critical current density is determined by the size, density, and pinning strength of the vortex traps.

Now suppose the matrix and precipitate separately exhibit magnetization curves of the type shown at the upper right in Figure 10. At medium fields there is a region where the precipitate particles are more diamagnetic than the matrix. In this instance the precipitate particles repel flux and thus cannot serve as pinning centers. The corresponding critical current density in this field range is very small as indicated at the lower right in Figure 10. On the other hand, at higher fields the matrix is the more diamagnetic, and the precipitate particles again act as traps. The result is a critical current curve with a peak at high fields as shown. Similar arguments involving materials with different T_c, H_c, and $\varkappa$ values have been invoked to account for critical current densities which *increase* with increasing temperature [62].

The situation in real inhomogeneous materials is certainly much more complex than that discussed above. Proximity effects are always present, and types and distributions of trapping centers may be established in infinite variety. Although great ingenuity has been exercised in attempts to optimize critical currents in supermagnet materials, it is clear that the practical limits have not yet been achieved. Some techniques which have been used with success should nonetheless be mentioned. Cold working produces dislocation tangles, and it also redistributes metallurgical phases. Heat treatments are used to introduce concentration gradients and to create mixtures of metallurgical phases. Fast neutron irradiation is used to produce damage sites. Incomplete sintering leaves voids which act as traps. Small ferromagnetic particles have even been used as traps [63]. The most impressive result yet reported, of the order of 2×10^6 A/cm^2 at 30 kG, was obtained in V_3Si which was first doped with U and then irradiated with neutrons, thereby producing a high density of fission damage sites having particularly good trapping characteristics [64]. An excellent discussion of the effects of metallurgical variables on superconducting properties is to be found in a review by Livingston and Schadler [65].

It should be emphasized at this point that vortex-state transport currents are metastable. At finite temperatures some vortices are unpinned by thermal activation. They migrate under the action of the Lorentz force until they are again trapped. This is a dissipative process known as "flux creep" [61, 66]. In the operating regime of a supermagnet, flux creep ordinarily takes place at a rate which is negligible for practical purposes. On the other hand, at very high fields and currents the Lorentz force exceeds the pinning force, and flux creep is replaced by highly viscous flux flow analogous to that already discussed for homogeneous materials.

A particularly interesting synthetic high-field superconductor has been studied by Bean and coworkers (see Bean [67]). They formed multiply connected filamentary superconducting structures by forcing Type I materials into porous Vycor glass. The small pore size, of the order of 100 Å, allowed superconductivity to survive to fields greater than 20 kG, where average current densities greater than 10^5 A/cm^2 could be supported.

Other interesting synthetic high-field superconductors have made use of Nb filaments. Moderately good high-field, high-current characteristics were obtained for Nb filaments formed in a *normal* Th-rich matrix by drawing of a eutectic mixture [68]. On the other hand, disappointing results were obtained for Nb filaments formed in a normal Cu matrix by wire bundling techniques. A composite wire was produced having within its cross section 10^7 Nb filaments each about 100 Å in diameter [69]. The poor characteristics obtained are most likely attributable to proximity effects enhanced by long electron mean free paths, a result of using high purity starting materials.

Occurrence of Superconductivity

Thus far we have dealt mainly in generalities. This has been possible because with very few parameters (e.g., T_c, H_c, and $\varkappa$) the theory can accommodate (at least approximately) a vast variety of material behavior. This fact has tended to lead workers in this field into hoping that *simple* correlations can be established for the occurrence of superconductivity. My own literature survey, admittedly incomplete, has uncovered about 30 criteria for the occurrence of superconductivity, one of which even relates the transition temperature to the boiling temperature [70]. Also among these criteria is the speculation due to Kapitza [71] that superconductivity should appear in all metals, a view avidly revived in recent years (see Matthias, Geballe, Compton, Corenzwit, and Hull [72] and subsequent discussion). In any event, with careful selection of points from among the more than 900 entries in the Roberts survey of 1964 [73], I suspect it is possible to make a case for almost any of these 30 criteria. The difficulty lies in trying to substitute a very simple criterion for what really must require a very complicated calculation on a very complicated system. As a one-time Fermiologist I subscribe to the view that it would be meaningless to attempt theoretically to predict, say, the electrical resistivity of a normal metal without a complete knowledge of the electronic energy states in the vicinity of the Fermi surface, the phonon spectrum, and the details of the coupling between the electrons and phonons. Similarly, without these ingredients a prediction of T_c is not possible. Even with them success is frustrated by mathematical complexities of the microscopic theory. Thus, although some of the simple empirical criteria for the occurrence of superconductivity have on occasion been very useful, their value has been mainly statistical.

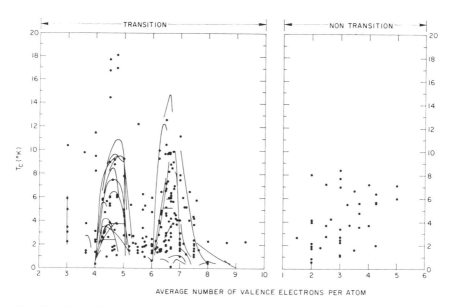

Fig. 11 Transition temperature versus average number of valence electrons per atom for all superconductors listed in the Roberts survey of 1961 [74] and for a few additional materials studied more recently. Ranges of alloy composition are represented by solid lines.

What is meant by "statistical" is shown in Figure 11. There T_c is plotted against the average number of valence electrons per atom for all superconductors listed in the Roberts survey of 1961 [74] and for a few additional materials studied more recently. Data for nontransition metals, their alloys and compounds appear on the right. On the left are data for transition metals, their alloys and compounds, and their compounds with nontransition metals. Ranges of alloy compositions are represented by solid lines. The peaks and high point densities in the vicinities of 3, 5 and 7 electrons per atom for the transition metals illustrate the Matthias regularities [75]. Such a plot indicates statistically where high transition temperatures will most likely be found, but it does not allow prediction of T_c for a particular material. (It could in fact delude us into ignoring interesting materials which lie off the peaks.) Since Figure 11 was prepared, the region below 2 electrons per atom, which seemed almost to be a forbidden one, has been populated by a number of Au compounds and alkali metal compounds [75–79]. The close approach of compounds like Au_5Ca and Au_5Ba to one electron per atom lends support to the speculation that superconductivity may ultimately be found in pure monovalent metals.

In any event, if plots like Figure 11 won't allow prediction of T_c, then where is guidance to be sought? A good starting point at least is the following well-known approximate result of the BCS theory.

$$T_c = 0.855\theta_D e^{-[1/N(O)V]}. \qquad (13)$$

Here θ_D is the Debye temperature, $N(O)$ is the electronic density of states at the Fermi level, and V is the effective electron–electron interaction parameter. Because the result (13) was obtained on the assumption of simplified universal forms for the electronic structure, the phonon structure, and the interaction parameter, we cannot use it to predict T_c. More rigorous applications of the BCS method to actual electron and phonon dispersions would be required. However, we can use (13) to gain an appreciation for what factors are of importance.

It has long been recognized that a large value of $N(O)$ favors superconductivity (e.g., see Daunt [80]). In fact, the Matthias' regularities in T_c are simply reflections of similar statistical peaking in the electronic density of states. It has also long been appreciated that superconductivity is favored by a large electron–phonon scattering contribution to the normal state electrical resistivity, and this simply corresponds to a large value for V. Observed dependences of T_c on isotopic mass are either in accord with the dependence on θ_D indicated in (13), or can at least be rationalized in terms of more realistic electronic and phonon structures (e.g., see Garland [81]).

For many years major attention was focused on attempts to increase $N(O)$ as a means of achieving superconductivity. V, which was generally thought to vary slowly across the periodic chart, was largely ignored. It was thought to be something about which very little could be done. However, it was recently pointed out by M. L. Cohen [82, 83] that the effective V should be greatly enhanced in materials characterized by a many-valley electronic structure (the more valleys the better), a large intervalley electron–phonon scattering probability, and a large dielectric constant. These conditions are apparently met in certain polar semiconductors. When such materials are made degenerate by suitable doping, superconductivity is observed, even for charge carrier concentrations 10^5 times smaller than in ordinary metals [84–88]. This remarkable success of Cohen's ideas serves to amplify the BCS approach and adds support to the belief that electron–electron interaction by the virtual exchange of phonons is the dominant mechanism for superconductive pairing.

Now let us consider the characteristic parameters for some of the typical and atypical superconducting materials as listed in Table 1. Pure W (see [89]) is an extreme Type I superconductor with a transition at 12 millidegrees and a thermodynamic critical field of only about 1G. The corresponding $\varkappa$ is nearly equal to zero. This results in pronounced supercooling effects which make detection of superconductivity more difficult and complicate considerably the determination of equilibrium parameters.

Pure Pb is a Type I(a) superconductor near T_c. At lower temperatures $\varkappa$ increases, and Pb becomes a Type I(b) superconductor with a sheath phase appearing above H_c [90].

Pure Nb is an "intrinsic" Type II superconductor (see [91]). The term "intrinsic" is used here to denote a pure material which is Type II solely on the basis of the parameters which determine the first term on the right in the expression (8) for $\varkappa$, the second term being negligible by virtue of a very large electronic mean free path (or equivalently a vanishingly small ρ_n).

Impure, inhomogeneous Nb_3Sn, with an upper critical field of the order of 200 kG, is one of the most important supermagnet materials. However, a perfect crystal of the ordered compound Nb_3Sn would most likely exhibit a much smaller upper critical field. An estimate of H_{c2} for such a hypothetical perfect crystal may be made using some of the parameters deduced for imperfect Nb_3Sn by Cody [92]. The smallest reasonable estimate, $H_{c2} = 28$ kG, is listed in Table 1 to illustrate the possible extreme influence of crystal perfection.

The disordered solid solution alloy Ti-16 at. % Mo exhibits a strongly *paramagnetic* vortex state [18–20] (Hake, private communication). The very large $\varkappa$ value, 67, stems mainly from the second term in (8), a consequence of a very short electronic mean free path. The very small coherence length, 56 Å, is also noteworthy.

The compound C_8K, which is formed by the addition of potassium to graphite, has a lamellar-type structure. The enormous anisotropy leads to upper transition fields which differ by a factor of ten for the applied field parallel and perpendicular to the layers [93].

Reduced $SrTiO_3$ is a degenerate semiconductor which exhibits Type II behavior [84, 94, 87]. The transition temperature is very strongly dependent on charge carrier concentration as was expected. The data in

TABLE 1
Superconducting Parameters for Various Material Types

	T_c (°K)	$H_c(O)$ (G)	$\kappa(T_c)$	$\lambda(O)$ (Å)	ξ_G (Å)	$H_{c1}(O)$ (G)	$H_{c2}(O)$ (G)
W extreme Type I	0.012	1.07	0.0028*	820*	290,000*		
Pb Type I(a) T_c Type I(b) 0°K	7.18	803	0.33 0.5	390	1,135*		
Nb Type II	9.25	1,993	0.781	350	430*	1,735	4,040
Nb$_3$Sn	18.3	5,300	3.0*	390*	125*	1,950*	28,000*
Ti-16 at. % Mo extreme Type II paramag. vortex state	4.2	890	67	5,000*	56*	<75	~64,000
C$_8$K $\quad H\parallel$ layer compound $\quad H\perp$	0.39						250 25

SUPERCONDUCTING MATERIALS

Material							
SrTiO$_{3-}$ 10^{20} cm^{-3} degen. semiconductor	0.38	20*				4	300
Ba$_{0.13}$ WO$_3$ purple and red	1.9						
La-0.8 at.% Gd T_c magnetic ordering T_m	1.5 / 0.45						
MoIr ordered	8.8						
MoIr disordered	1.85						
InSb metallic, metastable	1.89	100		300,000	1,000*		
Be amorphous, metastable	6.5		large				107,000*

* Indicates calculated values.

Table 1 correspond to approximately 10^{20} cm^{-3}. Particularly noteworthy is the enormous penetration depth, 300,000 Å, a consequence of the very small charge carrier concentration.

One of the tungsten bronzes, Ba$_{0.13}$WO$_3$, which has a purple to red color, has been included in Table 1 to illustrate that superconductors seem to occur in almost all colors. Although a considerable number of the tungsten bronzes have been found to be superconducting [95–98] there seems to be little information on superconducting parameters other than T_c.

In the dilute alloy La-0.8 at.% Gd, a localized magnetic state is associated with the Gd impurities, and the superconducting transition is depressed from the value 4.9°K characteristic of pure hcp La to 1.5°K. When this dilute alloy is cooled further, to about 0.5°K, the Gd moments order magnetically. There exist a number of such systems in which localized-moment ordering and superconductivity appear to coexist. (For pertinent references see Finnemore, Hopkins, and Palmer [99].)

The system MoIr is of interest as an example of a material which undergoes an order–disorder transformation with respect to atomic site occupation. The ordered and disordered forms exhibit very different transition temperatures as noted in Table 1 [100].

The compound InSb is ordinarily a semiconductor, but it becomes metallic when subjected to a pressure of 27,000 atmospheres. If this form is cooled to low temperatures, and the pressure is then released, the metallic form is retained. It is a Type I superconductor with a transition at 1.89°K [101–3].

Measurements at temperatures down to 0.06°K have revealed no evidence of superconductivity in bulk Be. However, Be films condensed at liquid helium temperatures possess a metastable quasi-amorphous structure which is superconducting at temperatures up to ~8°K. If such a film is annealed at temperatures greater than 30°K and less than 60°K another highly disordered structure is formed having a superconducting transition at 6.5°K. The high degree of disorder leads to a very short electron mean free path, which in turn leads to a large $\varkappa$ and a surprisingly large H_{c2}, of the order of 107 kG [104].

Theoretical Speculations

In this final section we mention briefly some of the more speculative advice offered by theoreticians on the topic of the occurrence of superconductivity. According to Parmenter [105] superconductivity can be induced at relatively high temperatures by application of very high current densities, of the order of 10^9 A/cm^2. The basic idea is that the phonon-induced electron–electron interaction is greatly enhanced when the electron drift velocity is comparable to the velocity of sound. A lot of metal has been vaporized in attempts to observe this effect [106, 107]. The possibility that high-current superconductivity may occur in ionic semiconductors has been investigated theoretically by Eagles [108]. This may lead to the vaporization of some polar semiconductors as well.

Jaccarino and Peter [109] have speculated that a transition to a superconducting state may take place at very high magnetic fields in ferromagnetic materials in which the effective exchange field seen by the conduction electrons opposes the applied field. The transition would presumably take place when the applied field is large enough to compensate the effective exchange field.

Fröhlich and Terreaux [110] have predicted that at very high magnetic fields a superconducting semimetal or degenerate semiconductor will undergo a further phase transition characterized by vanishing electrical resistivity in a single direction, the direction of the magnetic field. They conclude also that the effect will persist to very high temperatures.

It has been suggested by Selivanenko [111] that undoped semiconductors may be made superconducting by illumination with a laser source of sufficient intensity to populate the conduction bands.

Ginzburg and Kirzhnits [112] have suggested that "two-dimensional" superconductivity may exist at the surfaces of metals and semiconductors as a consequence of electronic surface states and microscopic coupling mechanisms which are peculiar to the surface. This kind of surface superconductivity is to be distinguished from the sheath phase of Saint-James and De Gennes. The latter is a high-field effect in a material for which the electronic states, phonon states, and coupling mechanisms are homogeneous throughout the volume.

According to Ginzburg [113, 114], dielectric and semiconducting overlays might be used to induce surface superconductivity with a very high transition temperature. In this case electrons would interact with one another via the polarization they produce in the dielectric overlay. Ginzburg and Kirzhnits [112] (1964) have also proposed that surface charging be used as a means for producing superconductivity at the surface of a dielectric. A review of the various proposed surface superconductivity mechanisms has been published by Silvert [115].

A particularly radical theory has been proposed by Little [116, 117]. He suggests that superconductivity may occur at room temperature or above in "one-dimensional" organic systems, specifically, long polymers with side chains. He has indicated how an interaction might exist between electrons in the central chain as a result of their mutual interaction with side chain electrons. Such a system would presumably be superconducting only along the central chain.

This speculation has been compounded by Atherton [118] who suggests that by application of an electric field transverse to the central chain it will be possible to switch superconductivity off and on. His estimates imply that units invoking this principle might be suitable as elements of a computer (or artificial brain) which would be electrically compatible with the nervous systems of biological organisms.

Although the speculative nature of the topics of this last section should be emphasized it should nonetheless be remembered that superconductivity itself is doubtless a more bizarre reality than speculations prior to its discovery might have allowed.

I wish to acknowledge with thanks a critical reading of this manuscript by R. R. Hake.

Note added in proof: Since this paper was written there have been a few pertinent developments. Essmann and Trauble [119] have observed the triangular vortex lattice structure of the vortex phase in a very direct fashion. They employed decoration methods analogous to the Bitter techniques used in studies of ferromagnetic domain structure. Further studies of the pressure-induced metastable metallic form of InSb [120] have suggested that this material may be a Type II superconductor rather than a Type I superconductor as indicated in [103]. Further investigations of Be [121, 122] suggest that the films described in [104] may have been contaminated with W. Thus, whether or not pure Be films condensed at low temperatures would have properties as discussed above is not known.

References

[1] Ginzburg, V. L., and Landau, L. D. 1950. *Zhur. Eksp. i Teoret. Fiz.* 20: 1064.

[2] Abrikosov, A. A. 1957. *Zhur. Eksp. i Teoret. Fiz.* 32: 1442 [translation: 1957. *Soviet Phys.—JETP* 5: 1174].

[3] Gor'kov, L. P. 1959. *Zhur. Eksp. i Teoret. Fiz.* 36: 1918 [translation: 1959. *Soviet Phys.—JETP* 9: 1364].

[4] Gor'kov, L. P. 1959. *Zhur. Eksp. i Teoret. Fiz.* 37: 1407 [translation: 1960. *Soviet Phys.—JETP* 10: 998].

[5] Saint-James, D., and de Gennes, P. G. 1963. *Physics Letters* 7: 306.

[6] Bardeen, J., Cooper, L. N., and Schrieffer, J. R. 1957. *Phys. Rev.* 108: 1175.

[7] Cribier, D., Jacrot, B., Farnoux, B., and Rao, L. M. 1966. *J. Appl. Phys.* 37: 952.

[8] Tomasch, W. J. 1964. *Physics Letters.* 9: 104.

[9] Tomasch, W. J. 1965. *Phys. Rev.* 139: A746.

[10] Cardona, M., and Rosenblum, B. 1964. *Physics Letters* 8: 308.
[11] Bon Mardion, G., Goodman, B. B., and Lacaze, A. 1965. *J. Phys. Chem. Solids* 26: 1143.
[12] Maki, K. 1964. *Physics* 1: 21.
[13] Maki, K. 1964. *Physics* 1: 127.
[14] Maki, K. 1966. *Phys. Rev.* 148: 362.
[15] Caroli, C., Cyrot, M., and de Gennes, P. G. 1966. *Solid State Comm.* 4: 17.
[16] Helfand, E., and Werthamer, N. R. 1966. *Phys. Rev.* 147: 288.
[17] Werthamer, N. R., Helfand, E., and Hohenberg, P. C. 1966. *Phys. Rev.* 147: 295.
[18] Hake, R. R. 1965. *Phys. Rev. Letters* 15: 865.
[19] Cape, J. A. 1966. *Phys. Rev.* 148: 257.
[20] Barnes, L. J., and Hake, R. R. 1966. *Annales Academiae Scientiarum Fennicae, Series A*, Vol. XI (210): 78.
[21] Kim, Y. B., Hempstead, C. F., and Strnad, A. R. 1965. *Phys. Rev.* 139: A1163.
[22] Stephen, M. J., and Bardeen, J. 1965. *Phys. Rev. Letters* 14: 112.
[23] Niessen, A. K., and Staas, F. A. 1965. *Physics Letters* 15: 26.
[24] Reed, W. A., Fawcett, D., and Kim, Y. B. 1965. *Phys. Rev. Letters* 14: 790.
[25] Staas, F. A., Niessen, A. K., and Druyvesteyn, W. F. 1965. *Physics Letters* 17: 231.
[26] Druyvesteyn, W. F., and Stass, F. A. 1965. *Physics Letters* 19: 262.
[27] Fiory, A. T., and Serin, B. 1966. *Phys. Rev. Letters* 16: 308.
[28] Otter, F. A., and Solomon, P. R. 1966. *Phys. Rev. Letters* 16: 681.
[29] Marcus, P. M. 1965. *Low Temperature Physics LT9*. New York: Plenum Press. P. 550.
[30] Fink, H. J. 1965. *Phys. Rev. Letters* 14: 853.
[31] Swartz, P. S., and Hart, H. R. 1965. *Phys. Rev.* 137: A818.
[32] Bean, C. P., and Livingston, J. D. 1964. *Phys. Rev. Letters* 12: 14.
[33] Joseph, A. S. and Tomasch, W. J. 1964. *Phys. Rev. Letters* 12: 219.
[34] DeBlois, R. W., and DeSorbo, W. 1964. *Phys. Rev. Letters* 12: 499.
[35] Sandiford, D. J., and Schweitzer, D. G. 1964. *Physics Letters* 13: 98.
[36] Abrikosov, A. A. 1964. *Zhur. Eksp. i Teoret. Fiz.* 47: 720 [translation: 1965. *Soviet Phys.—JETP* 20: 480].
[37] Fink, H. J., and Kessinger, R. D. 1965. *Phys. Rev.* 140: A1937.
[38] Fink, H. J., and Barnes, L. J. 1965. *Phys. Rev. Letters* 15: 792.
[39] Barnes, L. J., and Fink, H. J. 1966. *Physics Letters* 20: 583.
[40] Park, J. G. 1966. *Phys. Rev. Letters* 16: 1196.
[41] Boato, G., Gallinaro, G., and Rizzuto, C. 1965. *Solid State Comm.* 3: 173.

[42] Böbel, G., and Ratto, C. I. 1965. *Solid State Comm.* 3: 177.
[43] Tomasch, W. J. and Joseph, A. S. 1964. *Phys. Rev. Letters* 12: 148.
[44] Lynton, E. A. 1962. *Superconductivity*. London: Methuen and Co. Ltd.
[45] Tinkham, M. 1963. *Phys. Rev.* 129: 2413.
[46] Maki, K. 1965. *Ann. Phys.* 34: 363.
[47] Blatt, J. M., and Thompson, C. J. 1963. *Phys. Rev. Letters* 10: 332.
[48] Thompson, C. J. and Blatt, J. M. 1963. *Physics Letters* 5: 6.
[49] Falk, D. S. 1963. *Phys. Rev.* 132: 1576.
[50] Thompson, C. J. 1965. *J. Phys. Chem. Solids* 26: 1053.
[51] Singh, A. D. 1965. *Physics Letters* 16: 98.
[52] Tavger, B. A., and Kresin, W. S. 1966. *Physics Letters* 20: 595.
[53] Josephson, B. D. 1962. *Physics Letters* 1: 251.
[54] Josephson, B. D. 1965. *Advances in Physics* 14: 419.
[55] Anderson, P. W., and Rowell, J. M. 1963. *Phys. Rev. Letters* 10: 230.
[56] Hauser, J. J., Theuerer, H. C., and Werthamer, N. R. 1966. *Phys. Rev.* 142: 118.
[57] Minnigerode, G. v. 1966. *Z. Physik* 192: 379.
[58] Hauser, J. J., and Theuerer, H. C. 1965. *Physics Letters* 14: 270.
[59] Woolf, M. A., and Reif, F. 1965. *Phys. Rev.* 137: A557.
[60] Hempstead, C. F., and Kim, Y. B. 1964. *Phys. Rev. Letters* 12: 145.
[61] Anderson, P. W. 1962. *Phys. Rev. Letters* 9: 309.
[62] Livingston, J. D. 1966. *Bull. Am. Phys. Soc.* 11: 225.
[63] Alden, T. H., and Livingston, J. D. 1966. *Appl. Phys. Letters* 8: 6.
[64] Bean, C. P., Fleischer, R. L., Swartz, P. S., and Hart, H. R. 1966. Technical Report No. AFML-TR-65-431 (by General Electric Company for Air Force Materials Laboratory), 98.
[65] Livingston, J. D., and Schadler, H. W. 1964. *Progress in Materials Science* 12: 183.
[66] Kim, Y. B., Hempstead, C. F., and Strnad, A. R. 1962. *Phys. Rev. Letters* 9: 306.
[67] Bean, C. P. 1964. *Rev. Mod. Phys.* 36: 31.
[68] Cline, H. E., Rose, R. M., and Wulff, J. 1963. *J. Appl. Phys.* 34: 1771.
[69] Cline, H. E., Strauss, B. P., Rose, R. M., and Wulff, J. 1966. *J. Appl. Phys.* 37: 5.
[70] Schaaffs, W. 1963. *Naturwissenschaften* 50: 470.
[71] Kapitza, P. 1929. *Proc. Roy. Soc.* (London) A123: 342.
[72] Matthias, B. T., Geballe, T. H., Compton, V. B., Corenzwit, E., and Hull, G. W. 1964. *Rev. Mod. Phys.* 36: 155.
[73] Roberts, B. W. 1964. *Progress in Cryogenics*. London: Heywood and Company, Ltd. P. 161.

[74] Roberts, B. W. 1961. General Electric Research Laboratory Report No. 61-RL-2744M.
[75] Matthias, B. T. 1957. *Progress in Low Temperature Physics* II: 138.
[76] Matthias, B. T. 1959. *J. Phys. Chem. Solids* 10: 342.
[77] Merriam, M. F. 1962. *Bull. Am. Phys. Soc.* 7: 474.
[78] Arrhenius, G., Raub, Ch. J., Hamilton, D. C., and Matthias, B. T. 1963. *Phys. Rev. Letters* 11: 313.
[79] Hamilton, D. C., Raub, Ch. J., Matthias, B. T. Corenzwit, E., and Hull, G. W. 1965. *J. Phys. Chem. Solids* 26: 665.
[80] Daunt, J. G. 1950. *Phys. Rev.* 80: 911.
[81] Garland, J. W. 1967. *Phys. Rev.* 153: 460 and to be published.
[82] Cohen, M. L. 1964. *Phys. Rev.* 134: A511.
[83] Cohen, M. L. 1964. *Rev. Mod. Phys.* 36: 240.
[84] Hein, R. A., Gibson, J. W., Mazelsky, R., Miller, R. C., and Hulm, J. K. 1964. *Phys. Rev. Letters* 12: 320.
[85] Schooley, J. F., Hosler, W. R., and Cohen, M. L. 1964. *Phys. Rev. Letters* 12: 474.
[86] Frederikse, H. P. R., Schooley, J. F., Thurber, W. R., Pfeiffer, E., and Hosler, W. R. 1966. *Phys. Rev. Letters* 16: 579.
[87] Schooley, J. F., Colwell, J. H., and Ambler, E. 1966. *Bull. Am. Phys. Soc.* 11: 207.
[88] Pfeiffer, E. R., and Schooley, J. F. 1966. *Bull. Am. Phys. Soc.* 11: 208.
[89] Johnson, R. T., Vilches, O. E., Wheatley, J. C. and Gygax, S. 1966. *Phys. Rev. Letters* 16: 101.
[90] Cardona, M., and Rosenblum, B. 1965. *Low Temperature Physics LT9*. New York: Plenum Press. P. 560.
[91] Finnemore, D. K., Stromberg, T. F., and Swenson, C. A. 1966. *Phys. Rev.* 149: 231.
[92] Cody, G. D. 1964. *RCA Review* 25: 414.
[93] Hannay, N. B., Geballe, T. H., Matthias, B. T., Andres, K., Schmitt, P., and MacNair, D. 1965. *Phys. Rev. Letters* 14: 225.
[94] Ambler, E., Colwell, J. H., Hosler, W. R., and Schooley, J. F. 1966. *Phys. Rev.* 148: 280.
[95] Raub, Ch. J., Sweedler, A. R., Jensen, M. A., Broadston, S., and Matthias, B. T. 1964. *Phys. Rev. Letters* 13: 746.
[96] Sweedler, A. R., Raub, Ch. J., and Matthias, B. T. 1965. *Physics. Letters* 15: 108.
[97] Sweedler, A. R., Hulm, J. K., Matthias, B. T., and Geballe, T. H. 1965. *Physics Letters* 19: 82.
[98] Bierstedt, P. E., Bither, T. A., and Darnell, F. J. 1966. *Solid State Comm.* 4: 25.

[99] Finnemore, D. K., Hopkins, D. C., and Palmer, P. E. 1965. *Phys. Rev. Letters* 15: 891.
[100] Sadagopan, V., Pollard, E. R., Giessen, B. C., and Gatos, H. C. 1965. *Appl. Phys. Letters* 7: 73.
[101] Bömmel, H. E., Darnell, A. J., Libby, W. F., and Tittman, B. R. 1963. *Science* 139: 1301.
[102] Geller, S., McWhan, D. B., and Hull, G. W. 1963. *Science* 140: 62.
[103] Stromberg, T. F., and Swenson, C. A. 1964. *Phys. Rev.* 134: A21.
[104] Lazarev, G. G., Semenenko, E. E., and Sudovtsov, A. I. 1963. *Zhur. Eksp. i Teoret. Fiz.* 45: 391 [translation: 1964. *Soviet Phys.—JETP* 18: 270].
[105] Parmenter, R. H. 1965. *Phys. Rev.* 140: A1952.
[106] Gittleman, J. I., Bozowski, S., Parmenter, R. H., Rosenblum, B., Rosi, F. D., Seidel, T. E., and Wicklund, A. W. 1963. RCA Report on Contract No. DA-36-039-SC-88959, ARPA Order 210–61, Dept. of the Army Project No. 9800.
[107] Hartlin, E. M., Wertheimer, R. M., and Graham, G. M. 1964. *Can. J. Phys.* 42: 1282.
[108] Eagles, D. M. 1966. *Physics Letters* 20: 591.
[109] Jaccarino, V., and Peter, M. 1962. *Phys. Rev. Letters* 9: 290.
[110] Frölich, H., and Terreaux, C. 1965. *Proc. Phys. Soc. (London)* 86: 233.
[111] Selivanenko, A. S. 1964. *Fiz. Tverdogo Tela* 7: 1567 [translation: 1965. *Soviet Phys.–Solid State* 7: 1263].
[112] Ginzburg, V., L., and Kirzhnits, D. A. 1964. *Zhur. Eksp. i Teoret. Fiz.* 46: 397 [translation: 1964. *Soviet Phys.—JETP* 19: 269].
[113] Ginzburg, V. L. 1964. *Zhur. Eksp. i Teoret. Fiz.* 47: 2318 [translation: 1965. *Soviet Phys.—JETP* 20: 1549].
[114] Ginzburg, V. L. 1964. *Physics Letters* 13: 101.
[115] Silvert, W. 1966. *Physics* 2: 153.
[116] Little, W. A. 1964. *Phys. Rev.* 134: A1416.
[117] Little, W. A. 1965. *Scientific American* 212: 21.
[118] Atherton, D. L. 1965. *Nature* 205: 687.
[119] Essmann, U., and Trauble, H. 1967. *Physics Letters* 24A: 526.
[120] Dorer, G. L., and Joseph, A. S. (to be published).
[121] Olsen, C. E., Matthias, B. T., and Hill, H. H. 1966. *Bull. Am. Phys. Soc.* 11: 806.
[122] Glover, R. E., 1967. *Bull. Am. Phys. Soc.* 12: 76.

4
Quantum Engineering
James E. Mercereau

The title quantum engineering is intended to suggest an ideal or a philosophy rather than a promise that I can describe how to engineer anything, because by training I am a physicist. Perhaps the best thing I can do is to describe how we make certain gadgets and what those gadgets can do. The engineering task is much more difficult and perhaps is still impossible. What we are trying to do in our laboratory is to develop the technology from which such engineering can be done.

The philosophy—or the idea behind the philosophy—is simply that a superconductor is in a macroscopic quantum state; that idea is by now perhaps twenty years old. It was introduced by London [1] a long time ago. Here is a chunk of metal—something that you can hold in your hand—and yet it is a macroscopic quantum system. Of course, we can use quantum mechanics to describe anything if we are proficient enough with the arithmetic. What I mean by macroscopic quantum mechanics is that there is a single, fairly simple wave function that describes at least part of the electrons in the metal. The complex wave function ψ for a superconductor has as its modulus the square root of some density ρ, with a phase factor γ.

$$\psi = \sqrt{\rho} \cdot e^{i\gamma}.$$

In this big chunk of metal, some of the electrons can be described in terms of a quantity as simple as ψ. There is a number for the density and

The author is manager, Ford Scientific Laboratory, Newport Beach, California, and research associate, California Institute of Technology, Pasadena, California.

a number for the phase, and if we know those two numbers we can describe everything relating to a certain fraction of the electrons in this chunk of metal. We know, for example, that the time derivative of this wave function is simply proportional to the energy in the system; we can describe probability currents and if we calculate the probability current from ψ, we find that the current is proportional to the vector potential **A**.

$$\mathbf{j} = -\left(\frac{\rho e^2}{m}\right)\mathbf{A} = -\lambda^{-2}\mathbf{A}$$

If we have a macroscopic quantum system, this must always be true. This is one of the two London equations. It says that the current is proportional to the vector potential, which gives rise to a different kind of electrodynamics than you would expect, say, for a piece of copper, where the current is proportional to the electric field. As a matter of fact, this implies that the current accelerates proportionally to the electric field, and that, of course, implies that the resistance is identically zero. If we choose, we can combine this with a Maxwell equation that says that the curl of the magnetic field is equal to the current. If we do that, and if we say that the sources of the magnetic field are outside the material that we are talking about, we are forced to the conclusion that the magnetic field inside the material is zero except for a little depth, λ, at the surface, depending on the size of the constant of the proportionality that relates the current to the vector potential.

That is all very nice, and it seems to describe pretty well most of the materials that are called superconductors. The difficulty, however, is that it is completely useless, because the only thing described that is really worthwhile in an engineering or a technological sense is the fact that the resistance is zero. A great deal of effort has been devoted to using that property—superconducting magnets, perhaps transmission lines, high-Q cavities, etc; but it all revolves around one simple idea—the fact that the resistance is zero. As a matter of fact, there are two classes of superconductors, Type I and Type II. Type I, where the resistance is zero and the magnetic field is excluded, I have described. Type II is different in that the resistance is still zero but the magnetic field may be inside; this kind of material turns out to be more useful than Type I for the superconducting magnets. The magnetic field is inside and can, therefore, interact with the current, so we might think that we are in trouble—and as a matter of fact, we are. The current interacts with the field, pushes it through, and generates voltages. The reason the field sometimes does not move is simply that the superconducting material has impurities, strains, or whatever in it. So the useful properties of superconductors depend on the state of imperfection of the superconducting material to a large extent, at least for superconducting magnets. From an engineering point of view, that seems exactly backwards, so, rather than use the imperfections in the superconducting state to make useful things, we went back and tried to make use of the perfection of the superconducting state.

Let us go back to the point where we had to find two parameters—a density and a phase factor; from there we had derived an equation that related current to some electrodynamic variable. In the process, however, we managed to lose the phase factor and had only something that was proportional to the density of superconducting electrons. The idea was: could we go back to the fundamental description and make something useful that had to do with this phase factor, which is a property of the quantum mechanical system? That is a clever idea, of course, and it was not done a long time ago simply because there was no phase detector. There was nothing available, from Hewlett-Packard or anybody else, that measured the quantum mechanical phase; so that part of superconductivity was dormant for a long time simply because there was no mechanism to make the necessary measurements. The realization that allowed us to build essentially a quantum mechanical phase detector was a theoretical description by Josephson [2] of the outcome of having two superconductors separated by a little bit but still so close together that electrons can tunnel from one to the other. They do not travel in a real way—they sort of imagine their way across—and it develops that the current density J for the transfer from one superconductor to the other is not proportional to the vector potential, as the London equation would have you believe. It is proportional to an amplitude J_0 and to the sine of the phase difference $\Delta\gamma$ across this barrier, between the two superconductors,

$$J = J_0 \sin \Delta\gamma.$$

Now we have a current density, something that can be measured, that states, in fact, how the phases change, at least between two points. But there is still some trouble because it is only a current density, and, after all, we are interested in macroscopic currents—we would like to put wires and meters on it to find out what happens. So we really have to sum the current density over all the contact area across which the current is going to flow; but when we do that we have to remember that the phase difference $\Delta\gamma$ is not a gauge-invariant quantity; in fact, it depends on the magnetic field. We want to express it in a gauge-invariant way, and the easiest method is to express it as the phase difference between two points—a point a and a point b—and then subtract the part that makes it non-gauge-invariant, giving us

$$\Delta\gamma_{ab} = \Delta\gamma_0 - \frac{2e}{\hbar}\int_a^b \mathbf{A}\, dx.$$

We shall ignore the derivation, except to say that a detailed model enters in only in the amplitude J_0 and its temperature dependence. That the current depends on the sine of the phase difference is independent of any model used. It depends only on the fact that the time derivative of the wave function is proportional to the energy, so that no matter how you couple the two superconductors together, the current is going to depend sinusoidally on the phase difference. Josephson took the BCS theory and calculated an actual value for the current density that turned out to be correct. When we sum the current density over all the area of contact to get the total current I, we have to be careful when we have a (uniform) magnetic field. The vector potential is then a function of position because the curl of the vector potential is equal to the magnetic field. We have the normalized magnetic flux ϕ/ϕ_0 in the junction, contained in the argument of the sine, that will give us a diffraction pattern when we have summed over the area. The term describing the diffraction effect will be multiplied by $\sin \Delta\gamma_0$, where $\Delta\gamma_0$ is some residual phase difference.

$$I = I_0(\sin \Delta\gamma_0)\frac{\sin[\pi(\phi/\phi_0)]}{\pi(\phi/\phi_0)}$$

Remember that ϕ is the flux in the junction and $\phi_0 = h/2e$ the flux quantum.

This was first experimentally verified by Anderson and Rowell [3] at the Bell Telephone Laboratories. With a device such as this, if no current flows through it, the phase difference, $\Delta\gamma_0$, becomes equal to zero and everything is all right. As we turn the current on, the phase difference takes some value appropriate to that magnitude of current. But still we have nothing to measure, and we are almost back where we were. We will have something to measure, at least so far as our description goes, only when we destroy this description—when we try to pass so much current through the device that the description is no longer adequate. Under those circumstances, if the current cannot go through by the mechanism described it has to go some other way, so it starts to tunnel in the usual Giaever-type tunneling, and a voltage appears. A voltage will appear whenever this $\sin \Delta\gamma_0$ reaches its maximum value, which is one, and the maximum current that can be passed through for a zero voltage is given by

$$\frac{\sin \pi(\phi/\phi_0)}{\pi(\phi/\phi_0)}.$$

Figure 1 shows essentially the way the current goes with voltage, which I have already described.

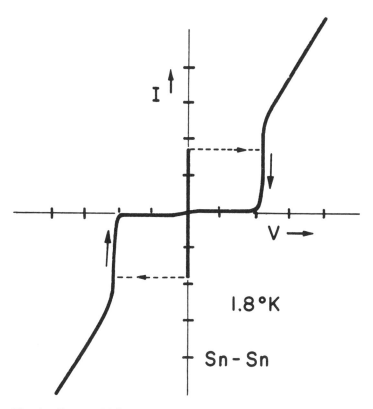

Fig. 1 Current–Voltage characteristic of a tin–oxide–tin Josephson tunnel junction.

I am describing a particular way of instrumenting these devices, and before we finish I will mention several other techniques. If we apply a current source to the junction, of course the current starts up from zero and no voltage appears because it is possible to get current flow without a voltage, as we have just seen. The relative phase is changed, and the phase angle is determined by the magnitude of current. At the point where the sine of the phase difference can get no bigger, the device reverts to the normal state and a voltage appears. This magnitude—this maximum current at zero voltage—has a diffraction-like character associated with the amount of magnetic field enclosed in the junction, as we have said before, and that behavior is shown in Figure 2. The junctions are very simple to make: take one piece of a superconductor, let the surface get a little dirty with oxygen, for example, and put another piece of superconductor on top of it. Figure 2 shows the maximum current that can be put through the resulting structure as a function of the magnetic field.

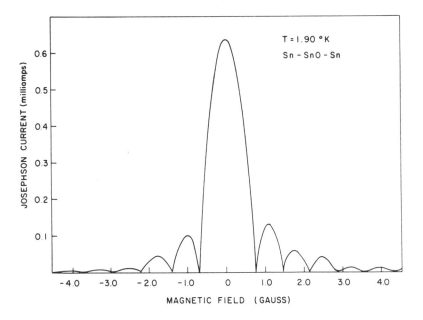

Fig. 2 Maximum Josephson current as a function of magnetic field showing the characteristic diffraction-like behavior.

This proves that the phase detector works, but the trouble again is that in making the measurement we have destroyed a great deal of the required information because we are making measurements only when the current has reached its maximum. In fact, we are making measurements not in a pure superconducting state; we are making measurements just after we have been driven away from that state. Ideally, we would like to make our measurements in such a way that we preserve some of the phase information that we have destroyed. One way of doing this is to compare the behavior of one of these junctions with another one and to compare the behavior of the phase differences between two points. An object designed for this purpose is called an interferometer, of course.

The interferometer is simple: put two of these devices in parallel and then compare the phase $\Delta\gamma_1$ across one with the phase $\Delta\gamma_2$ across the other. If we hook them together with copper wires, naturally $\sin\Delta\gamma_1$ goes to its maximum value; $\sin\Delta\gamma_2$ goes to its maximum value; and we get twice as much maximum current as we did with one. Again, this is not very useful, but if we had hooked them together with pieces of superconductor, the idea of a macroscopic quantum system tells us that if we know the phase at point 1 we can find out what it is at point 2 by adding to the phase at the first point essentially the distance divided by the wavelength multiplied by 2π. Also, the phase may change in time, so we have to add to it $\Delta\omega\, dt$, where $\Delta\omega$ is the difference in frequency between points 1 and 2.

$$\gamma_2 = \gamma_1 + 2\pi \int_1^2 \frac{dx}{\lambda} + \int \Delta\omega\, dt.$$

Using this prescription to relate the various phases ($\Delta\gamma_1$, $\Delta\gamma_2$), then making some calculations, we shall have two terms. One, of course, will be the line integral all around the circuit and the other will be the difference in relative phase. If we assume that the current through one junction is the same as through the other, which is the easiest method, we shall finally arrive at the total current that flows in transmission. (I am now referring to the total current in transmission, not the circulating current.) It is given by the sine of the integral of difference in frequency across the junctions, plus the term involving the wavelength. If we are to believe that quantum mechanics applies in a simple way, then one over the wavelength is the momentum divided by h, and $\Delta\omega = \Delta E/\hbar$. The difference in energy is prescribed as being due only to a voltage, so that it is $2eV$, although energy differences can arise in other ways. In this way, we get

$$I = I_0 \cos\left(\frac{1}{\hbar}\int 2eV\, dt + \frac{1}{\hbar}\oint P\, dx\right).$$

We now have a prescription for the operation of this quantum mechanical device. The line integral is the integral of the momentum, P, around the loop, and the other term inside the cosine, which depends on $2e/h$, is the integral of the voltage difference plus some arbitrary phase factor. This is how the transmitted current would flow in a device that is essentially a macroscopic quantum device; the behavior of the device is entirely determined by this prescription. A detailed model of superconductivity comes in only in the amplitude I_0; the rest of it has to be independent of any model of superconductivity. To be still more explicit, we shall describe some of our experiments that verify this prescription.

The momentum we are discussing is, of course, a mass times a velocity, but since the particles are charged we have to add the vector potential contribution. Since we are dealing with electron pairs, these quantities all involve two *m*'s and two *e*'s. The cosine term will therefore contain $2e/\hbar$ times the line integral of the vector potential, which, of course, is the amount of magnetic flux enclosed now in the large area, plus a term, the mechanical momentum, mv, also integrated around the loop, which is something like the mechanical angular momentum of the electrons. Thus, from an engineering point of view, we have a device that responds to magnetic flux (field times an area), to its state of rotation, and to the voltage,

$$I = I_0 \sin\left(\frac{2e}{\hbar} \int V\,dt + \text{constant}\right) \cos\left(\frac{2e}{\hbar} \oint A\cdot dx + \frac{2m}{\hbar} \oint v\cdot dx\right).$$

Our first task a few years ago was to show that, in fact, this prescription was adequate to describe what we saw.

The third figure shows how we fabricated the device, at least initially. We wanted to connect two pieces of superconductor together in such a

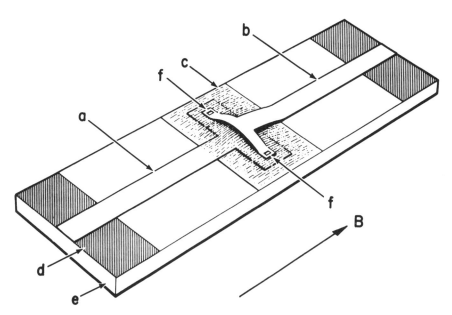

Fig. 3 Thin film superconducting interferometer. Films (a) and (b) are separated by a paint layer (c) except at oxide contacts (f). Magnetic field (B) is usually applied parallel to the glass substrate (c). Current is introduced at contacts (d).

way that they almost touched in two spots, so we evaporated a T-shaped film on a substrate. We painted over this lower superconducting film except for the two spots f; we let these lie around for a while to get dirty; and then we evaporated another piece on top. The result was a piece of superconductor connected to another piece of superconductor with both almost touching at those two spots. The two pieces of superconductor were separated by an area depending upon the thickness of the paint.

The first experiments [4] were, again, performed essentially to measure the maximum zero voltage current and to obtain the diffraction effect that we saw before. If, however, we apply a magnetic field perpendicular to the loop, we shall also get a flux enclosed in this area between the two superconductors that will cause the interference effect. The kind of data obtained are shown in Figure 4, which shows two interference patterns for two different structures such as these. It shows essentially only the central diffraction maximum of the diffraction pattern, but that maximum is then modulated by the interference effect. Curve A was for a fairly small area of flux, curve B for a larger area, and, of course, since the area gets bigger, the period of magnetic field gets smaller. If we measure the area as accurately as we can to determine what the period is, the result is $h/2e$, as expected. For all the materials tested (niobium, vanadium, lead, indium, aluminum, tin, and perhaps a few others), the periodicity is always $h/2e$. The amplitude of the current that you can pass through the device does depend on the temperature, but the

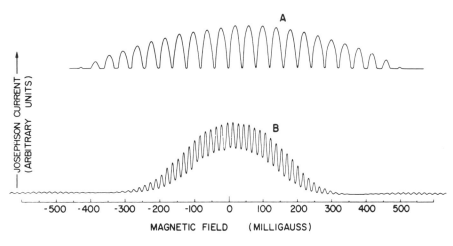

Fig. 4 Maximum Josephson current through a superconducting interferometer as a function of magnetic field. The difference in field periodicity between curves A and B is as expected for the differing areas.

periodicity shown on the curves is independent of the temperature. These experiments essentially verified our ideas.

We also did some measurements to show the effect of the rotation of the device. In this experiment the field was fixed and the device was rotated; thus, we achieved effects equivalent to those shown in magnetic field. The periodicity in this case is $h/2m$ rather than $h/2e$, but it is, again, adequately described by our equation.

We shall not go through the arithmetic, but we can also write an expression for the circulating current in an entirely similar way. The circulating currents are easier to measure at higher frequencies. To examine the voltage effect, we did the experiment at microwave frequencies where the voltages get to be large enough to be manageable. We looked at the circulating current in a device as such I have described, by examining the impedance of the circuit. Figure 5 shows a suitable circuit. The circuit is very simple—just a thin film evaporated on some substrate, but we have scratched it in such a way that there are narrow sections. In the one shown there are two; we have also done the experiment with only one. We can use as many scratches as we want. It breaks down at the weakest one and works there. A structure such as this, when put in a microwave cavity, allows measurement of the circulating currents through the impedance of the device. Figure 5 essentially shows the variation of the microwave impedance of that circuit as a function of magnetic field: again, it is periodic, and, for this particular area, the period is about 15 microgauss. This information is a chart recording of the output from the cavity after crystal rectification. There was no attempt to clean up the signal.

So far, all we have shown is the effect of $\oint A \cdot dr$. At microwave frequencies the presence of the term $\int V\, dt$ allows these measurements to be made even in the presence of a voltage. To verify that this really occurs, the amplitude of the effect should be demonstrated to be a function of voltage. The corresponding term in the current, of course, is the sine of the integral of a voltage. If the voltage itself depends on time, the result is a frequency modulation effect with side bands, and the current amplitude, as a function of voltage, should vary approximately as a Bessel function; that is, it should first wiggle, and then at higher voltages the effect should vanish. Omitting the analysis, Figure 6 shows that this is, in fact, true. The behavior is not that of a simple Bessel function, but the effects are roughly so describable: the period agrees with that of a J_1 Bessel function, but the amplitude, of course, does not. There is evidence of other effects. We cannot describe this as nicely as we should like, but we are not concerned, because we have used a description involving the behavior of a perfect Josephson junction. In fact, the experiment was done in a superconducting circuit that had

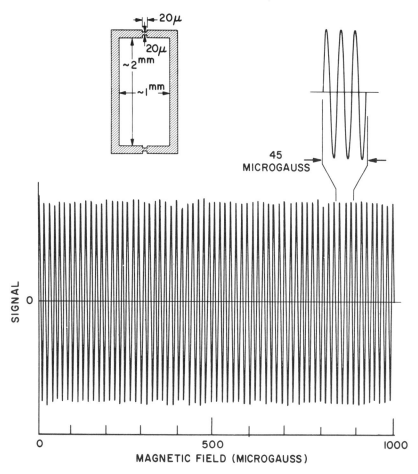

Fig. 5 Continuous thin film superconducting interferometer. The variation of signal with magnetic field is a measure of the microwave impedance variation of this circuit.

no junction at all—simply little, narrow sections. These narrow sections are being driven into what might be called a flux flow by the current that passes through them. Since there is a critical current density, the superconductor becomes critical in those narrow sections before it becomes critical in the rest of the circuit, which provides the barrier across which you get flux flow. In fact, that is the way we do all the experiments now. We no longer try to make dielectric barriers, superconducting junctions. We simply operate with pieces of superconductor that are sculptured in such a way that we have two large pieces of superconductor separated

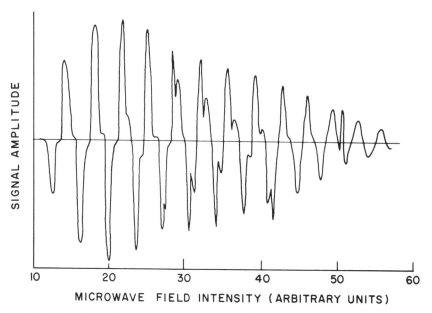

Fig. 6 Signal amplitude from the continuous film interferometer as a function of applied microwave field intensity. Such variation is typical of these interferometers and is suggestive of the expected Bessel function dependence.

by a very tiny piece, and we drive a current through the structure that puts the tiny piece into the critical, or flux flow, state that allows such effects to occur.

All this background is intended to show that the description we have is adequate for the phenomena that we are talking about. What we want to do now is to engineer some of these devices into useful instruments. I can describe some of the instruments that we have and some of the work we are doing with them. Recall that such devices are sensitive to the state of rotation, to the magnetic field, to the magnetic flux, and to the voltage. The easiest to instrument is a magnetic field; so we decided to make a few magnetometers as our first attempt at instrumentation. Figure 7 shows the kind of information that we are working with. The periodicity is in the current, and the amplitude of the current is shown as a function of the magnetic field applied to the device. It turns out to be a sinusoidal magnetic field, and so the sweep is linear in time. The period changes with position, but if you relate it to magnetic field, it corresponds to 10^{-7} gauss from peak to peak. Figure 7 contains information taken directly from the interferometer and fed through wide-band

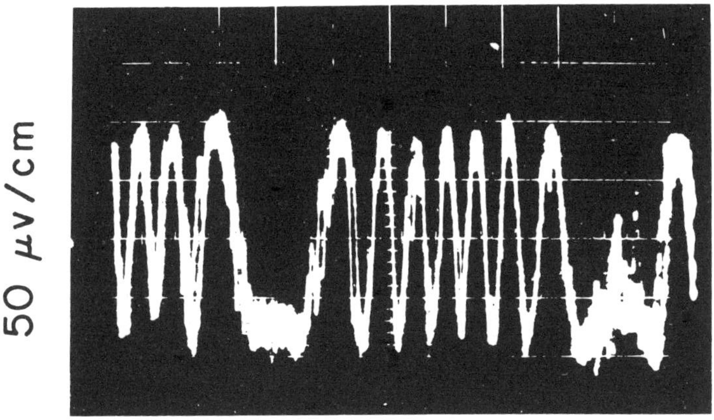

QUANTUM ENGINEERING

Fig. 7 Maximum Josephson current as a function of magnetic field for a relatively sensitive interferometer. This interferometer is used in the magnetometer shown in Figure 8.

Fig. 8 Superconducting magnetometer and associated electronics built by Forgacs and Warnick of the Ford Scientific Laboratories.

amplifiers into an oscilloscope; there has been no attempt to make the results look nice.

One way to make a magnetometer [5] from such a sensor is to use a feedback loop to keep the field constant. This feedback circuit drives a current through a field coil to compensate for external field perturbations. The instrument then senses the control current. We have succeeded in building an instrument with a noise level of less than 10^{-9} gauss for a response time of 0.1 sec.

75

We built another instrument which, again, is not an absolute instrument. It is essentially a digital fluxmeter, as it measures changes in magnetic flux in numbers of quanta. Its sensitivity is about 10^{-9} gauss, and its response time is about 10^4 quanta per second. Figure 8 shows one of these instruments. The noise that we see at these levels seems to be due either to an imperfection in the shielding that we cannot eliminate, even with superconducting shields or to noise in the electronics. There are situations in which the superconducting sensor at these noise levels is not the source of the noise. Consequently we cannot say what the ultimate sensitivity would be, only what we have obtained thus far. The random fluctuations in the earth's field are 10^{-5} gauss. There is also the diurnal variation, which is about a milligauss per day. Such a magnetometer used in research would have to be well shielded.

References

[1] London, F. 1950. *Superfluids*. New York: John Wiley and Sons, Inc.
[2] Josephson, B. 1962. *Physics Letters* 1: 251.
[3] Anderson, P. W., and Rowell, J. 1963. *Phys. Rev. Letters* 10: 230. *Phys. Rev. Letters* 12: 159.
[4] Jaklevic, R. C., Lambe, J., Silver, A. H., and Mercereau, J. E., 1964. *Phys. Rev. Letters* 12: 159.
[5] Forgacs, R. L., and Warnick, A. 1966. *IEEE Natl. Convention Record*, pt. 10 (March): 90–99; 1967. R.S.I. 38: 214.

5
Superconducting Lines for the Transmission of Large Amounts of Electrical Power over Great Distances

Richard L. Garwin and J. Matisoo

Introduction

Before World War II, there was little incentive for the electric utility companies to develop means of long-distance transmission of electrical energy. In fact, the only high-voltage long lines in existence were those required to transport power from hydroelectric sources to population centers. Electrical energy transmission over long distances was avoided, mostly because of the high cost of transmission over lines (a compromise between power loss in a small conductor and the capital expense of a large-diameter line). It was more economical to build generating stations near the major consumption centers and to transport the energy to the generating stations in some other way; for example, as coal by barge or railway. This procedure was also consistent with the structure of the electric utility industry, which not long ago consisted of many comparatively small local companies, each serving its own area, quite independently of one another.

The recent (and future projected) growth of the utility industry has

R. L. Garwin is with the IBM Watson Laboratory, Columbia University, New York, N. Y. He was formerly with the IBM Thomas J. Watson Research Center, Yorktown Heights, N. Y.
J. Matisoo is with the IBM Thomas J. Watson Research Center, Yorktown Heights, N. Y.

Reprinted from *Proc. IEEE* 55 (April, 1967): 538–48. This article is a revised version of the paper presented by the authors at the Conference.

forced a re-examination of the economics of past practice. In particular, although the costs of coal transportation by railway have been decreasing, this cost is still substantial[1]. Furthermore, the utilities have recently become aware of the advantages of power pooling. By tying together formerly independent power systems they can save in reserve capacity (particularly if the systems are in different regions of the country), because peak loads, for example, occur at different times of day, or in different seasons. To take advantage of these possible economies, facilities must exist for the transmission of very large blocks of electrical energy over long distances at reasonable cost.

Other problems which face the utility industry also require for their solution economical means of electrical energy transmission. Among these are generating-station site location and full utilization of existing or of novel power sources. The location of fossil-fuel plants near or in high-population-density areas has disadvantages. Suitable sites may be unavailable or very expensive; air pollution or thermal pollution problems may limit the generating capacity below optimum size; there is still much resistance to placing nuclear power plants in congested areas, because of possible dangers, no matter how improbable they may be. Distant sites make available economies of scale which are particularly important in nuclear plants.

Should it be possible to transmit large amounts of electrical energy with negligible loss, fossil-fuel as well as nuclear plants could be placed so as to offer no hazard to urban areas, and the choice of location could be made entirely on economic considerations; in particular, mine-mouth operation of steam plants would lead to large savings in coal transportation costs.

Economical transmission would also make possible serious consideration of alternative sources of energy; for example, one might build large solar generators in deserts or tap sources of hydroelectric power much more distant than those now considered practical for development.

In this paper we consider the problem of economical electrical energy transmission; in particular, the problem of transmitting very large blocks (100 GW) over long distances (1000 km). These numbers, while large, are not unreasonably so, since interties of 4-GW capacity over such a distance are currently under serious consideration [3]. Large-block

---

[1] The average mine cost of coal is $4.50 per ton (see Howard [1]), whereas the average transportation cost (1962) was $3.37 per ton [1]. Since approximately 1.2 kWh of electrical energy is obtained per pound of coal, the transportation cost of coal is approximately 1.4 mills/kWh. This transportation cost is an appreciable fraction of busbar energy costs (1964) of 6.8 mills/kWh (Olmsted [2]).

transmission is especially important since, whereas satisfactory conventional solutions exist for the transmission of multi-megawatt blocks over distances of ~ 500 km, this is not the case, as will be made clear in this paper, for 100 GW over 1000 km.

There are, as in every engineering problem, a number of possible alternative solutions, the choice among which must be made on the basis of cost or return. The first alternative, of course, is not to transmit electrical energy at all but rather to ship coal. The shipment of coal does offer some advantages; for example, easy storage of energy near the consumer, which eliminates, or at least minimizes, the peak-load problem on the transmission line and provides security against short-term (few-week) interruptions of the transportation system.

The transportation cost of coal for a power transmission task of this magnitude, however, exceeds $1 billion per year. This cost is large; therefore, considerable effort is justified to reduce it.

Another alternative is conventional (probably dc) long-distance EHV transmission by ordinary metallic conductors. This solution has the advantage of the use of existing and proven technology. On the other hand, this, too, is expensive. Lower bounds for cost involving such a line are $670 million for the conductor, perhaps several times this amount for towers and, most important, a line loss of $\sim \$340$ million per year. To these we must add converter costs, if the transmission is to be at dc (Appendix A).

We propose as a solution a superconducting line; i.e., a transmission system using superconductors to carry current. The superconductors would, of course, be refrigerated so that their resistivity is zero. The major power loss then would be the power required to maintain the line at $4°K$.

In what follows we shall sketch the design of such a line.

Details of the Proposed Transmission Line

The proposed transmission line consists of two insulated superconducting cables which are maintained at $4°K$ (the boiling point of liquid helium at one atmosphere), a convenient working temperature.

The power transported down the line is still, of course, proportional to the voltage difference between the two cables, and to the current flowing (with essentially zero dissipation) in the cables.

The design problem can be divided into three rather natural, but obviously very much interdependent parts: the superconducting cable itself, the refrigeration system, and the transmission line as a whole, including its relationship to existing facilities.

The cable problem involves: the choice of an appropriate superconductor and of cable dimensions based on the necessary currents and voltages; demonstration of the need for dc transmission, which is required by the economics of the refrigeration system and has far-reaching consequences for the line as a whole; finally, some details of construction.

The refrigeration system uses the principles of the common Dewar flask (Thermos bottle); i.e., the primary coolant is liquid helium (He), the insulation is vacuum, and the radiant heat shield is cooled by liquid nitrogen (N_2). The power to run the refrigerators is tapped from the line.

Finally, there are the problems of providing for redundancy and repair of the transmission line and the tie-in problems to ac systems at ordinary temperature.

A. Superconducting Cable

As the superconducting material from which the cable is made, we choose niobium-tin (Nb_3Sn). At temperatures low compared with its "critical temperature" ($T_c = 18°K$), this material remains superconducting in a field of 100,000 G [2] while carrying current densities of 200,000 A/cm² [4]. (For a brief summary of properties of superconductors, see Appendix B.) If we choose the transmission voltage as 200 kV, the total current must be 0.5×10^6 A. Limiting the current density to 10^5 A/cm² implies a superconductor cross section of 5 cm². This choice insures that the self-magnetic field is below the critical one of 10^5 G, even for cable made entirely of superconductor; if the cable is "stabilized"[6] by the inclusion of low-resistivity aluminum or copper, which occupies space and thus increases the radius, the magnetic field will, then, be still smaller.

The choice of transmission voltage of 200 kV is obviously conservative, since 345-kV conventional insulated underground lines are in use [7]. It is probable, since insulation problems are much eased at low temperatures, that the transmission voltage could be raised to 500 kV, thereby achieving a 2.5-fold increase in the power transmission capability of the line, or alternatively a 2.5-fold reduction in the current and thus in the cross section (and cost) of the superconductor.

Nearly all transmission lines in the world today transmit alternating current. The major reason is the ease with which the voltage can be changed ("transformed") from link to link along the system. Even so,

---

[2] Nb_3Sn has been shown to remain superconducting in fields of 220,000 G [5].

there are reasons to prefer dc transmission under some circumstances. Insulation requirements are less severe with dc transmission, and only two conductors are required per dc circuit rather than the three required in the ac case. Furthermore, the maximum practical length of ac cable is limited by the capacitive "charging" current, which, of course, does not exist on dc lines.

There are, however, far more compelling reasons for preferring dc transmission on *superconducting* lines.

Alternating currents exert alternating forces on flux lines in the superconductor, causing irreversible motion of the flux lines (thus, loss). This produces heat. In fact, as we shall see below, normal heat leak into the line is $\sim 4 \times 10^{-4}$ W/cm. Hysteresis losses, if transmission were at 60 Hz, would exceed this by a factor or 10^8, making line refrigeration impossible. [For details, see Appendix C, (1) through (3).]

A recent paper [21] demonstrates the impracticability of ac transmission on refrigerated lines. Note, however, that the coolant circulation in Figure 3 of that paper ("twin-duct jacket") is unsatisfactory because such a counter-current flow does not remove appreciable heat.

Even with a (nominally) dc line, fluctuating demand and particularly the initial loading of the line cause dissipation, largely as a result of hysteresis in flux motion. This loss is a serious design consideration if the line is not to serve only as a base supply, at constant load.

Fluctuating demand can readily be handled. Equation (4) (in Appendix C) gives the dissipation due to fluctuations as

$$H = 2 \times 10^{-15}(\Delta I)^3 F \quad \text{W/cm} \tag{4}$$

where F and ΔI are the frequency and amplitude of the fluctuation, respectively. Thus, the switching on and off of individual customers of 1 MW or 5 A every millisecond produces a dissipation of only 2.5×10^{-10} W/cm.[3]

Turn-on is a more difficult problem. Equation (2) (in Appendix C) gives a total dissipation of 0.07 W/cm for current growing from zero to 0.5×10^6 A in one hour, exceeding the radiative heat leak by a factor of 200!

Electrical loading of the line begins after the line has been cooled to 4°K from room temperature, a process which requires approximately five days. In terms of such a time scale a turn-on time of one or even two days is not unreasonable. The dissipation for a one-day turn-on time will exceed normal heat leak by a factor of ten.

---

[3] Note that if the cable is coated with copper, losses due to fluctuating demand, etc., will be eddy-current losses in the copper. The corresponding loss is 2×10^{-6} W/cm.

In any case, turn-on dissipation may be reduced by appropriate cable design. Thus, dissipation is reduced by a factor of g if the superconducting cable is indeed made of fine wires each of diameter g times smaller than the cable diameter. [A g of 200 equals the heat leak for a one-hour turn-on interval, and a g of 10, the heat leak for a one-day turn-on interval. A 3-cm cable diameter means an elementary diameter of 1.5×10^{-2} cm (or 3×10^{-1} cm for $g = 10$).]

The wires in a cable may be insulated from one another and effectively transposed over a 1-to-10-km length so that the wires initially at the core of the cable are on the outside for a similar distance, or the wires may be of shorter length in a normal-conducting (copper) matrix, the length of the wires being a compromise between steady-state loss and turn-on loss.

Clearly, there is room for considerable engineering in cable design. Since no lines of such capacity have yet been fabricated, the exact nature of the line depends upon the actual engineering compromises found to be necessary. The principle of the multistrand cable, however, is sound, as shown by the successful use of such cable in superconducting magnets [8].

Anelastic losses in ac transmission would also be large. (They would exceed normal heat leak by a factor of 10^4.) Because of the alternating current and, therefore, magnetic field, the pressure exerted on the conductor by the magnetic field varies between zero and its maximum value, 120 times per second. Since the material is not perfectly elastic, some of this energy is lost as heat each cycle. An anelasticity of less than 10^{-5} would be required to have an elastic loss comparable with the heat leak. Such low anelasticity would be difficult to obtain (Appendix C). Furthermore, fatigue would also preclude the use of alternating current.

Another reason for dc transmission becomes clear when we realize that the characteristic impedance of the line is ~ 100 ohms, whereas the current voltage limitations require a load of about $0.4 \, \Omega$. The 100-Ω characteristic impedance implies that to carry 0.5×10^6 amperes on a long line requires a potential difference of 0.5×10^8 volts between cables and not 2×10^5 volts as in the dc case.

Unfortunately, superconducting cable is not produced or shipped in infinite lengths. This means that it is necessary to consider how to make joints. It will be difficult to make superconducting joints by the common methods since ordinary soft superconductors would be quenched anyhow in fields of 10^5 gauss. Therefore, we estimate the dissipation caused by pure normal-metal solder joints.

A cable cross section of 5 cm^2 means a mass of ~ 100 g/cm or 10 tons/km. Since 10 tons is a reasonable shipping weight, we must consider joints every km. A metal-film 10^{-1}-cm thick ($\rho = 10^{-8} \, \Omega \cdot$cm at 4°K) between butted superconducting sections 5 cm² in area will present a resistance of 10^{-12} ohms and will thus cause one-watt dissipation per joint. This might be handled without excessive temperature rise ($\sim 0.5°$K), but the joint can in addition be skived (the ends cut at 6° angle to the axis of the bundle). The surface of the film is thus increased by a factor of ten, which results in a tenfold reduction of dissipation, or the film thickness can then be allowed to increase to 10^{-2} cm to maintain one-watt dissipation per joint. It may be necessary to bleed liquid helium from the supply directly to the joints to keep them at 4°K. Since superconducting joints of 10^{-9}-Ω resistance have been made in high-field superconducting magnets, the normal joints considered here may be too conservative.

It is also necessary to consider the steady electromagnetic forces acting on the conductors. These forces are large. They can be approximated (at 0.5×10^6 A) by a uniform external pressure on each cable of 400 atm (6,000 psi) plus a force of repulsion of 10^9 dyn/cm (10^6 newtons per meter). Since the strength of plastics at low temperature is 10^{10} dyn/cm² or more, this force of repulsion can be readily supported by allowing the plastic-insulated cables to squeeze against the enclosing 4°K shell, which itself could be made of aluminum alloy or stainless steel some 2 mm in wall thickness. This wall thickness is sufficient to withstand the force.

B. Refrigeration System

A major problem of a superconducting transmission line is the refrigeration system. We shall discuss the design evolved and the reasons for the choices made in this section, and present the detailed calculation of heat leaks, the required refrigeration capacity, the flow rates of the cryogenic fluids, and the necessary vacuum pumping capacity in Appendix D.

Figure 1 shows a section of the entire transmission line and particularly the cooling system. The portion of the line which is at 4°K is interior to the 4°K vacuum wall. This pipe contains the liquid He line, as shown, and the two plastic-insulated superconducting cables. These cables may be pulled into the pipe loosely or they may be supported occasionally or continuously. The remainder of the space inside this 4°K wall serves as the He gas return (at one atmosphere).

Surrounding this wall is the 77°K radiation shield, which is cooled by the liquid (and gas) N_2 line; this, in turn, together with the gaseous N_2 return line (at 10-atm pressure) is soldered to the sheet-metal radiation shield. The shield is slit at the top so that it may be easily separated, should access to the inside become necessary. The shield-pipe assembly might be made in fifty-foot sections and the sections then soldered together, or the shield could be formed from a "continuous" coil at the installation site.

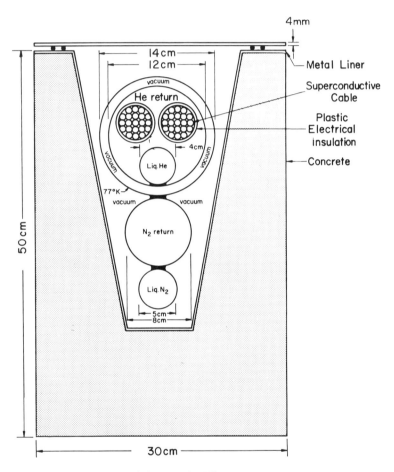

Fig. 1 Cross section of the 100-GW line.

The entire transmission line rests on polyethylene, nylon, or other cheap plastic bridges, which are spaced approximately three meters apart. These supports contribute negligible heat leaks.

Finally, there is the vacuum-tight metal trough in a concrete channel which forms the basic unit of support. The concrete channel, which will probably be fabricated at the site, can be either below or above ground.

The metal (tin-plated steel, or stainless steel, or perhaps aluminum) is glued with epoxy resin to the concrete. (The metal liner or a foil insert therein acts as an additional heat shield to radiation.) The metal sections of the trough are soldered together (or bonded with epoxy resin and the joints smoothed) so that they are vacuum-tight. To seal the trough, a continuous coil of aluminum (4 mm thick in 5-km sections which weigh 15 tons) is laid on top. Rubber gaskets and O-rings are used to make it vacuum-tight. This cover may be tipped up locally for inspection and repair. It has periodic apertures for vacuum pumping lines.

Looking now along the length of the line, the various elements of the cooling system and their spacing are shown in Figure 2. Every 20 km

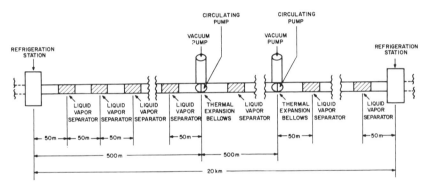

Fig. 2 A 20-km module of the 1,000-km, 100-GW line.

along the line there are refrigeration stations which supply liquid He and liquid N_2. The cryogenic fluids are pumped along and flow through their respective pipes. It is not a minor problem, however, to cool 20 km of line from a central source. The difficulty arises because the cryogenic fluid vaporizes as it flows along the pipe. The vapor thus produced reduces the density of the mixed phase and increases the velocity required to maintain the same mass-flow rate, thus necessitating large pumping power. Indeed, even with increased vapor density at low temperatures, the effect is still prohibitive at both He and N_2 temperatures.

We solve this problem by use of liquid-vapor separators at 50-m intervals; i.e., floats which vent the gas to the gas-return manifold, while retaining liquid in the supply pipe. The liquid-vapor separators at 50-m intervals also avoid the usual problem of choking on initial cooldown of the line [9].

The separators must be capable of handling an evaporation rate of ~ 1 cm^3 of liquid He per second (or 10 cm^3/s of gas) and ~ 3 cm^3/s of

N_2 liquid (or ~ 60 cm^3/s of gas). On cooldown, however, the entire liquid flow rate of the line must be allowed through the separator; clearly, it must be a device whose average position varies considerably but only slowly with demand. Separator design will not be considered further, except for noting that, if necessary, separators may include floats, switches, and motors and that they should fail-safe by closing.

Every 500 meters there is a propeller-type booster circulatory pump in the liquid He line. These are necessary because flow friction would produce a pressure drop of more than one atmosphere over the 20-km distance between refrigeration stations (at the design flow velocity of some 30 cm/s). With the booster pumps, the required pressure rise is ~ 0.10 atmosphere at a design flow rate of 0.5 liter per second.

Because of the longitudinal contraction of the refrigeration system (piping, lines, etc.) on cooldown, it is necessary to insert thermal-contraction bellows 1.5 m long every 500 m. (Thermal contraction problems are discussed in greater detail in Appendix E.)

The remaining major components of the line are the vacuum pumping stations which are spaced every 500 m along the length of the line. The vacuum pumping problem is really twofold. It is necessary to pump out the line on cooldown in a reasonable length of time (a few hours), which requires comparatively high-speed pumps (5 liters per second every 500 m), i.e., mechanical pumps. However, once the line is cooled the pumping requirements change. The cold He line is an excellent trap for all gases except He, so that only a low-speed ion pump is needed to pump out any small amount of He which may leak in (Appendix D). Indeed, the mechanical pumps may be replaced by liquid nitrogen-cooled zeolite traps, which will pump out the air which initially fills the trough.

Finally, the line as designed has a heat leak of 4×10^{-4} W/cm into the $4°$K section, which requires a refrigeration plant using motor power totaling 0.3 MW per 20-km section. The heat leak into the nitrogen shield will be 10^{-1} W/cm and requires a refrigeration motor capacity of 1.2 MW/20 km.

These are capacities required to keep the line at $4°$K. To cool the line initially in a reasonable length of time, additional capacity is required. Thus, we propose a permanently installed He refrigeration motor capacity of 1.5 MW and, if necessary, the trucking in of additional nitrogen refrigeration for cooldown. Thus, at each of the 20-km refrigera-stations we have a total installed motor power of ~ 3 MW.

To run the refrigerators, we tap power from the line. To provide 3 MW of power, we must plan to draw 15 amperes dc from the line at each 20-km station. The use of 200-kV dc motors is possible but certainly undesirable. Thus, the problem is to draw 15 amperes and convert it with high efficiency and low cost to 60-Hz ac. Since only some 200 MW total is needed for all refrigeration along the line (0.1 per cent of ac-dc-ac conversion load for the *transmitted* power), the cost of conversion for refrigeration is negligible, even in comparison with the cost of the conversion equipment for the transmitted power.

Each tap itself is a source of ~ 0.15-watt heat leak into the line for which no special provision need be made.

C. Interconnection and Reliability

We shall discuss in this section the interface problems caused by the low temperature and the necessary direct-current transmission of the line. Since the line must be connected to the room-temperature alternating-current power network, we must consider how this interconnection is to be made and how to solve the problems which arise.

One of the problems involves the heat leak at the ends of the line. Wires of optimum size extending from $4°K$ to $77°K$ dissipate and conduct $\sim 10^{-2}$ watts to the helium when one ampere is being removed and returned on a pair of leads [10]. (Heavier wires conduct more heat to the cold line; thinner wires dissipate more heat.) Since 0.5×10^6 amperes are being withdrawn, the minimum heat leak is ~ 5 kW. The additional refrigeration required will cost us an investment of ~ 1.5 MW of refrigeration motor. The expense is not large, but the cooling must be good enough to prevent the heating of the superconductor and the resulting spread of normal phase down the line. There are several ways to overcome this problem. The cable can be unstranded and the junction spread over 10-m length. This will reduce the cooling required to 5 W/cm, which seems feasible on a width of ~ 1 meter. There is no doubt that this can be done at reasonable cost. The current density in each of the strands is $< 10^5$ A/cm^2 and the field will be $\ll 10^5$ Oe; in any case, much below the critical values of Nb_3Sn.

Another problem is conversion of ac to dc at the source end and of dc to ac at the receiving end of the line. The detailed manner of rectification and inversion depends upon the cost of alternatives. In any case, it is likely that 6-phase or possibly even 12-phase rectification will be used, primarily to keep the ripple low. In fact, if dissipation due to hysteresis[4] is to be less than 10 per cent of the normal heat leak, we must have, in a solid superconductor, a ripple current less than 450 A. With 6-phase rectification the ripple voltage is 8×10^3 volts (rms) [11], so that the nonresonant current into $\sim 100\ \Omega$ is 80 A (rms). With 12-phase rectification it is only 20 A (rms); thus the inductance of the line itself serves to reduce the ripple current. Depending upon cost, various

---

[4] In copper-covered cable the losses are the eddy-current losses in copper. For losses to be 10 per cent of heat leak, ripple current must be less than 50 A.

compensating current or rectification schemes can be used to reduce the ripple to a negligible value.

The rectifiers (and inverters) themselves are likely to be mercury-arc pool type [12] or possibly silicon rectifiers and silicon-controlled rectifiers.

Finally, since it is foolhardy to assume perfect reliability, it would be wise to have a second line nearby with frequent interconnections between the lines as protection against power interruption. Connections are expensive, so relatively few are desirable. On the other hand, if no interconnections are made, there may be simultaneous breakdown of both lines. We have not definitely resolved this question; the problem is to be able to isolate a defective section of line so that it may be warmed to room temperature and repaired, which requires that there be no voltage across the lines. Too much refrigeration capacity would be required to bring the line to room temperature at every 20-km station at all times.

One solution would involve the construction of a "conventional" 200-kV, 100-GW dc line either underground or above ground near the superconducting line, with potential interconnections every 20 or 60 km. The normal line might be designed to dissipate, say, 20 per cent of the transmitted power in a *single* 20-*km* run, for the few days required to repair a fault in a single 20-km module of the superconducting line. Routing around a fault might involve shutting off the power for a few seconds to allow automatic unsoldering of the offending section, making contact with the parallel 20-km section of normal aluminum line (thus incurring a heat leak of some 5 kw at each end of the damaged section) and switching on power as demanded, up to the remaining 80 per cent of the superconducting line capacity. The switches required are not operable under load. The individual aluminum conductors required for the normal line are some 150 cm^2 in area, and the whole normal line may cost some $80 million for conductors—a factor 10 or more less than a normal line designed to carry the full 100 GW continuously for the full 1,000 km.[5] It is clear that there is still room for optimization of this emergency line, its tap spacings, etc. One use of the emergency line is to supply refrigerator and vacuum pump power initially while the superconducting line is still normal.

These are hazards involved in lines carrying such large currents, although they need not be serious. The energy in the line's magnetic field

$$W = \frac{1}{2} LI^2 = \int \frac{B^2}{8\pi} dv$$

[5] F. A. Otter, Jr., suggested (private communication, May 24, 1966) that this normal line might be made cheaply in this way.

is that stored in 1.4×10^5 gauss and 1×10^9 cm^3 or 10^{11} joules; that is, approximately the energy in 20 tons of TNT or in two tons of gasoline. In case of a ground fault, one can probably do much better than to isolate the line and to let the energy dissipate in the fault, thus gently blowing up a few feet of line. Even a dead short-circuit across a perfectly-conducting line will receive only $2 \text{ V}/z_0 = (4 \times 10^5)/10^2 = 4{,}000$ A until the reflected waves are received from the terminals. If one short-circuits the line at some point, the current will thus rise in steps of 4,000 A at a mean rate of 400,000 A/s. The resultant arc can hardly be called catastrophic. Indeed, even if the line conductors themselves are open-circuited, the flashover will short-circuit the line and there will be time for a switch at each refrigerator station to throw $\sim 50\ \Omega$ across the line. This will drain the energy in 0.1 second. Since this places 2×10^9 joules into each resistor, the temperature of each 10-ton cast-iron resistor will rise to 500°C.

D. Cost Estimate (Zeroth Approximation)

Here we try to estimate the capital and operating costs of one of the superconducting transmission lines described above. We shall not include in the cost estimate such items as the cost of right-of-way acquisition and clearing (although this cost is likely to be an order of magnitude less than that for conventional overhead line), cost of on-site labor, etc. In addition, we do not include the cost of transformers at the terminals, since these are required on any conventional line, no matter how short.

A last item which must be considered part of the capital cost of the superconducting line is the converters. There are no single converters currently in existence which are capable of handling the currents and voltages required. There are, however, mercury-pool converters which will safely handle 100 kV and a few thousand amperes [13]. If we extrapolate from the available cost figures of $1/kW, the total cost for two stations (one at each end) is $200 million, which is not very high. (See Table 1.)

Although the above costs may be in error by a significant factor, it is clear that the major cost (apart from converters) is the cost of the superconductor. There are various schemes by which the critical current density of superconductors might be increased. One such scheme is to fabricate synthetic high-field, high-current superconductors with small superconducting filaments in a suitable normal metal matrix [14]. The matrix-metal-to-superconductor-volume ratio, as well as the filament diameter and the materials, have to be appropriately chosen; otherwise no high-field properties are obtained [15]. Another scheme has been

TABLE 1

Item	Estimated Cost (millions)	Basis for Estimate
Superconductor	$550	10^4 tons of Nb_3Sn at $26/lb*
Refrigeration along line	$25	$0.5 million per He refrigerator of 1 kW at $4°K$ every 20 km†
End refrigeration	$5	These must handle a total of 10 kW (2×5 kW) or an equivalent of 10 1-kW refrigerators at $0.5 million each
Vacuum pumps	$1	Assumed $500 per pumping station
Fabricated metal	$20	Assumed cost of $1/lb and weight of 100 g/cm
Concrete	$5	Assumed cost of $10/yd^3 with total volume of $\frac{1}{2}$ yd$^2 \times 10^3$ km
Converters	$200	Cost of mercury-pool rectifiers $\sim$$1/kW‡ and of silicon rectifiers and silicon-controlled rectifiers $\sim$$1 kW§
Total	$806 (million)	

* Ultimate cost of Nb_3Sn estimated from Nb cost of $36/lb (*Chemical and Engineering News*, p. 65, February 3, 1964) and Sn cost of $2/lb (*Yearbook of the American Bureau of Metal Statistics for 1962*, New York, 1963 p. 132).

† Informal quotation to IBM Corporation by Arthur D. Little, Inc., Cambridge, Mass., February 1966.

‡ Informal quotation to IBM Corporation by General Electric Company on 3-MW mercury-pool rectifier (20 kV, 150 A).

§ Informal quotation to IBM Corporation by International Rectifier Corporation on silicon rectifier [1.2 kV, 550 A (rms)] and silicon-controlled-rectifier [1.2 kV, 400 A (rms)].

demonstrated by Bean *et al.* [16]. They have shown that current densities greater than 10^6 A/cm^2 (with 10^7 A/cm^2 conceivable) can be obtained (in V_3Si) by introduction of defects by internal fission in the superconductor. If, for example, the critical current density could be increased, to 10^7 A/cm^2, a hundred-fold reduction in the cost of the superconductor would result.

The operating costs arise primarily from the energy required to liquefy He, N_2 and the power required to run the vacuum pumps. In normal operation the total He refrigerator-motor power is 0.3 MW × 50 or 15 MW. Therefore, at a busbar cost of 6.8 mil/kWh, the He refrigerator operating cost is ~$1 million per year. The nitrogen cost is approximately $4 million per year, since the refrigerator capacity for the nitrogen line is roughly four times that of the He line. Thus, the operating costs are approximately $5 million per year, which, of course, compares very favorably with transmission losses on an ordinary line of this capacity.

Summary and Conclusions

We summarize the design characteristics of the line as follows.

Power capacity	100 GW (10^{11} W)
Voltage (dc)	200 kV (2×10^5 V)
Current (dc)	0.5×10^6 A
Line temperature	4.2°K (liquid helium)
Radiation shield	77°K (liquid nitrogen)
Length of line	1,000 km
Refrigerator spacing	20 km
Gas-liquid separator spacing	50 m
Booster pump spacing	500 m
Vacuum pump spacing	500 m
Thermal expansion bellows 1.5 m long (superconductors wound helically) spacing	500 m
Fraction of power dissipated in line and leads	$< 10^{-7}$
Fraction of power used for refrigeration	$< 10^{-3}$

We have offered another solution to the problem of economical electrical energy transmission, by sketching a design for a large-capacity, long-distance superconducting line and estimating the capital and operating costs for such a line. If our cost estimates are not too much in error, it is clear that the most economical solution to the over-simplified power transmission problem posed in the Introduction is a superconducting line of the general design described. This becomes particularly apparent when annual costs are examined. Thus, coal transportation cost is approximately $1 billion a year, ordinary EHV transmission losses ~$340 million a year, while the superconducting-line "losses" amount to only ~$5 million a year. Even the capital costs may favor the superconducting line over conventional EHV. The capital investment in EHV transmission is ~$1 to $1.5 billion, whereas the superconducting line cost is ~$606 million. (The converter costs are the same for EHV dc and superconducting line and therefore have not been included in the comparison.)

The superconducting line has essentially fixed annual operating costs; i.e., the refrigeration cost is almost independent of the current-carrying capacity of the line. Also, the capital costs associated with the refrigeration system are the same regardless of line capacity (assuming fixed 4.2°K operating temperature). What does scale is the superconductor cost (and converter cost), which varies directly with the power capacity of the line. With ordinary EHV transmission, as the power capacity is reduced there comes a point at which ac transmission becomes practical (eliminating converters). Losses then scale with power level, as does the capital investment in the line.

Thus, for sufficiently low power levels and over sufficiently short transmission distances, it will undoubtedly be more economical to use conventional ac EHV transmission. However, there will exist a power level and distance beyond which superconducting lines will prove more economical. Clearly, detailed engineering design and cost analysis is necessary to determine exact crossover points. We have shown, however, that circumstances may be such as to favor the novel approach of a superconducting power line.

Appendix A

Conventional EHV Transmission

Extra-high-voltage (EHV) transmission is presently carried out at 345 and 500 kV with, however, comparatively low capacity (say 1.2 GW).[6] If 100 GW were to be transmitted in the conventional way, the transmission must be dc and converter costs would be the same as in the superconducting case.

To carry the current one could use two aluminum (for the sake of argument consider aluminum rather than ACSR) conductors ~ 1200 cm^2 each in area [17] and at much higher voltage. The resistance of such a line (assuming 60°C operating temperature) is ~ 0.3 Ω, which results in ~ 5 per cent dissipation. To carry 100 GW a distance of 10^3 km at 5 per cent dissipation would require a current of 1.4×10^5 A at 750 kV.

Since there exist no detailed cost estimates in this power range, we obtain a rough lower limit to the costs involved as follows.

The weight of the aluminum conductors is $\sim 8 \times 10^5$ tons, which at $0.25/lb [18] for aluminum (or $0.38/lb for steel-reinforced cable) is a cost of $440 million for the material or some $670 million for the conductors. The cost of poles, etc., is probably several times this amount for a total capital investment of $\sim$$1 billion.

[6] A 2-GW, 266-kV *underground* dc system some 85 km long is under contract to be built into London (*New York Times*, April 30, 1966).

The main point, however, is that the line is extracting an annual toll of 5 per cent on the electricity transported, or a loss of $\sim 5 \times 10^{10}$ kWh per year at an annual cost of \$340 million.

Appendix B

Summary of Properties of Superconductors Pertinent to Current-Carrying Capacity

Superconductors are materials which below a temperature T_c (called the transition temperature) have zero electrical resistivity and exclude magnetic fields. This is true as long as the current density within the superconductor remains below a critical value and as long as applied magnetic fields are sufficiently small. Thus, provided these critical variables (temperature, current density, and magnetic field) have sufficiently low values, the resistance of the superconductor is zero. There are two kinds of superconductor, called Type I and Type II.[7] They are distinguished by their rather different magnetic and current carrying behavior (see Figure 3).

Weak magnetic fields are excluded from the interior of both Type I and Type II superconductors (except from a small surface layer, typically on the order of 500 Å). As the externally applied magnetic field is increased, in the case of Type I superconductors, the field penetrates completely at the thermodynamic critical field, and at higher fields the behavior is much the same as that of any normal metal.

In Type II materials, field penetration begins below the thermodynamic critical field (although this field may be much higher than the corresponding field in Type I materials). The field penetration continues with increasing field, to fields as high as 200 kG in Nb_3Sn. It is extremely important to note that a large fraction of the superconductor remains superconducting in high fields, although the magnetic field, on a macroscopic scale, penetrates uniformly [Figure 3(a)].

Another crucial difference between Type I and Type II superconductors is their current-carrying behavior. In Type I any transport current is carried on the surface (a consequence of flux exclusion), and, therefore, the total current-carrying capacity increases only as the diameter of the conductor. On the other hand, in Type II superconductors, in a sufficiently high field so that the field has penetrated, the

---

[7] For a good review of Type II superconductors see, for example, J. E. Kunzler, "High-field superconductivity," *Materials Research and Standards*, pp. 161–71, April 1965.

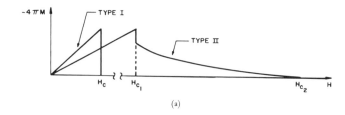

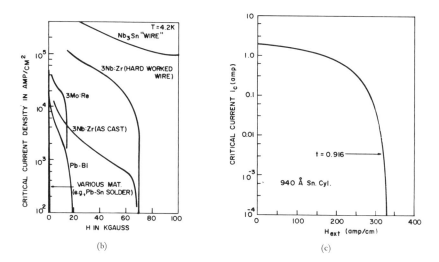

Fig. 3 (a) Magnetization vs. field for Type I and Type II superconductors (b) Critical current density vs. field for various Type II superconductors. [After J. E. Kunzler, High-field superconductivity, *Materials Research and Standards*, pp. 161–71, April 1965.] (c) Critical current vs. field for a Type I superconductor (a cylindrical thin film sample of tin, 940 Å thick). [After J. Mydosh and H. Meissner, *Phys. Rev.*, vol. 140, p. A1574, 1965.]

current density is essentially uniform over the cross section and thus the total current is proportional to the square of the diameter. It should be pointed out that the achievement of high-field high-current capability in these superconductors actually depends critically upon the existence of defects in the material which act as pinning sites to flux lines, and prevent their motion under the influence of Lorentz force ($J \times B$). When it does occur, the motion of flux lines produces a voltage drop along the length of the superconductor and thus effectively introduces resistance.

Appendix C

Losses in Alternating-Current Transmission

Hysteresis Losses It would be convenient to transmit conventional 60-Hz ac, for numerous reasons, such as ease in tapping power, and so forth. This is not possible for several reasons, two main ones being hysteresis losses in Type II superconductors (in fields greater than H_{c_1}; in practice, fields will always be greater than H_{c_1}), and anelastic losses produced by the changing magnetic field.

The hysteresis losses are due to the irreversible motion of flux under alternating-current conditions. In recent work Bean et al. [19] have shown experimentally that hysteretic loss (which is constant per cycle) is indeed the loss mechanism at power frequencies.

The hysteresis loss in a two-conductor superconducting cable carrying a large alternating current of amplitude I_0 is (Bean [19], p. 19)

$$H = 4 \times 10^{-9} I_0^2 F \quad \text{W/cm} \tag{1}$$

where F is the frequency in Hz. Thus, if transmission were to be 60 Hz the hysteresis loss would be approximately 6×10^4 W/cm!

Under noncyclic conditions there is also loss. Thus, we must consider the energy loss on energizing of the line. Bean [19] calculates this energy dissipation (per cable) as

$$0.5 \times 10^{-9} I_0^2 \quad \text{J/cm}$$

or a power loss of

$$H = 10^{-9} I_0^2 \tau^{-1} \quad \text{W/cm} \tag{2}$$

where τ is the line-energizing time, i.e., the time to bring the line-current (dc) from zero to a value I_0. For a τ of one hour, the loss is 0.07 W/cm; for a τ of one day, however, the loss is 3×10^{-3} W/cm.

Finally, because the current will not be ripple-free (converter ripple and fluctuating demand), we need to know the loss for small-amplitude alternating currents (ΔI).[8] Again Bean [19] gives this as

$$H = \frac{4 \times 10^{-10} (\Delta I)^3 F}{J_c R^2} \quad \text{W/cm} \tag{3}$$

where J_c is the critical current of the superconductor and R is the radius of the cable. For $J_c = 10^5$ A

$$H = 2 \times 10^{-15} (\Delta I)^3 F \quad \text{W/cm}. \tag{4}$$

---

[8] Note that if the cable is stabilized by the inclusion of a copper sheath, these ripple currents will flow in the copper sheath and the losses will be normal eddy-current losses.

Anelastic Losses The magnetic field squeezes the wire with pressure, $P(=B^2/8\pi)$. For a wire of modulus of elasticity Y, the stored energy is

$$\epsilon = \frac{P^2}{2Y}\pi R^2 \text{ erg/cm}. \tag{5}$$

In most plastics and even metals at ordinary temperature only one-half of this energy is recovered; i.e., the anelasticity a is 12. At low temperatures a will be smaller unless there is slippage.

$$\epsilon \simeq \frac{(4/\sqrt{2} \times 10^8)^2}{2 \times 10^{12}} 5 \simeq 2.5 \times 10^5 \text{ erg/cm}.$$

Since the stress goes to zero 120 times per second, the dissipation per centimeter is

$$h = 3 \times 10^7 \, a \text{ erg/cm·sec} = 3 \, a \text{ W/cm}. \tag{6}$$

To have this loss comparable with the heat leak requires that $a < 10^{-5}$.

These problems are avoided by dc transmission, as is the hysteresis loss in hard superconductors.

Appendix D

Refrigeration System

The heat transport into the line will be primarily by radiation. (Heat influx from 300°K to 77°K by conduction through nylon bridges of approximately 4 cm² in contact area and 8 cm long is roughly 10^{-3} W/cm or 1 per cent of radiative leak; from 77°K to 4°K for 1 cm² in contact area and 2 cm in length is approximately 10^{-5} W/cm again for nylon spacers 3 m apart, assuming a $\bar{\lambda} = 10^{-4}$ W/cm · °K over this temperature range.) The radiation per unit time per square centimeter between two surfaces of emissivity ϵ is given by

$$\frac{\epsilon}{2}\sigma(T_1^4 - T_2^4)$$

where T_1 and T_2 are the temperatures of the surfaces and σ is Stefan's constant (5.67×10^{-12} W/cm²·deg⁴). An emissivity of 0.05 is easy to achieve.

A radiation shield at 77°K is desirable (and conventional) to interrupt radiant heat at this temperature (T_2) and allow it to be rejected at room temperature (T_1) by expenditure of W units of work (Carnot cycle)

$$W = \frac{T_1 - T_2}{T_2}\frac{H}{E} \tag{7}$$

where

H = heat influx
E = refrigerator efficiency; i.e., fraction of ideal thermodynamic efficiency (~ 0.5 at $77°K$ but only ~ 0.2 at $4°K$).

The shield diameter is ~ 14 cm, so the heat influx into the nitrogen shield is $\sim 10^{-1}$ W/cm, and the heat flux from $77°K$ to $4°K$ (diameter ~ 12 cm, ϵ still 0.05) is 4×10^{-4} W/cm.

Therefore, over a 20-km length there is 2×10^5 W to be rejected from $77°K$ and 800 W from $4°K$. The corresponding refrigerator motor capacity must thus be 1.2 MW for $77°K$ and 0.3 MW for $4°K$ (working to $300°K$). This motor capacity is required to keep the line at $4°K$. However, ~ 1.5 MW of installed $4°K$ refrigerator motor capacity is necessary to cool the line in several days from $77°K$ to $4°K$.

It will require an even longer period of time to cool to $77°K$ with available capacity unless portable refrigerators to increase the total capacity are trucked in.

We now discuss the refrigeration problem in detail, and consider a pipe of radius R carrying fluid of density ρ at temperature T. A steady distributed heat flow, h erg/cm·s boils away liquid which is vented continuously into the return manifold. Consider the radius R independent of distance x, and the over-all length of the line as L. Denote the flow velocity as v. The velocity v, of course, varies along the length of the pipe and is zero at the far end ($x = L$). The volume of liquid transported per unit time is $V(x) = \pi R^2 v(x)$ and

$$\frac{dV(x)}{dx} = -\frac{1}{\lambda}\left[h + \pi R^2 v \frac{\rho v^2}{2} \frac{1}{100R}\right] \quad (8)$$

where λ is the heat of vaporization per cubic centimeter.

The first term on the right is the heat leak and the second term is the friction work. Since the Reynolds number for He in this (smooth) pipe is 5×10^5, the flow is clearly turbulent and the friction coefficient is ~ 0.0033 (velocity ~ 30 cm/s). Therefore, the kinetic energy of flow is lost as heat approximately every 50 diameters, and the second term of (8) represents this approximation.

On this basis we derive upper limits on the interval between refrigerators, and choose a distance well within the feasible range. Equation (8) can be written

$$-\lambda \pi R^2 \frac{dv}{dx} = h + \frac{\pi R \rho v^3}{200} \quad (9)$$

or

$$-dx\frac{h}{\pi R^2 \lambda} = \frac{dv}{1 + \frac{\pi R \rho v^3}{200h}}. \tag{10}$$

Defining $\pi R\rho/200h = \beta^{-3}$, the integral of (10) becomes

$$\frac{hL}{\pi R^2 \lambda} = \beta/\sqrt{3}\pi/6 + \beta/3\left[\frac{1}{2}\ln\frac{[\beta + v(0)]^2}{\beta^2 - \beta v(0) + v^2(0)} + \sqrt{3}\tan^{-1}\frac{2v(0) - \beta}{\beta\sqrt{3}}\right]. \tag{11}$$

The limiting length (as $v(0) \to \infty$) is

$$L_{\max} = \frac{\pi R^2 \lambda}{h}\frac{2\pi}{3}\beta/\sqrt{3}. \tag{12}$$

A practical maximum, however, is $\frac{1}{2}L_{\max}$ or, say,

$$L = \frac{R^2 \lambda \beta}{h} = 4\lambda R^{5/3}\rho^{-1/3}h^{-2/3}.$$

The corresponding initial fluid velocity is $\sim\frac{1}{3}\beta$.

Consider the N_2 problem. We have calculated $h \sim 10^6$ erg/s·cm, $\rho \sim 1$ g/cm³, and $\lambda = 1.6 \times 10^9$ erg/cm³. Therefore, for the 5-cm diameter N_2 pipe $L_{N_2} < 30$ km. For He refrigeration $h \sim 4 \times 10^3$ erg/cm·s, $\rho = 0.125$ g/cm³. For the 4-cm diameter He pipe we then have $L_{\text{He}} < 28$ km. Thus we might well choose a refrigerator spacing of 20 km.

The refrigeration rate is $\pi R^2 v\lambda = \pi R^2 \lambda(\frac{1}{3}\beta)$ or approximately 1 kW at 4°K. Larger flow rates can be obtained during cooldown from the same refrigerator.

Little energy is lost in allowing the N_2 vapor to warm to 300°K on its return to the refrigerator, since the latent heat of vaporization is roughly equal to the energy required to cool to the boiling point.

With helium, however, precisely the opposite is true, since the latent heat per cubic centimeter is smaller by a factor ~ 60; i.e., most of the energy required to liquefy helium gas goes into cooling the gas to 4°K.

The N_2 evaporation rate is ~ 2 liters per second in 20 km, or a gas flow rate at STP of 3,000 cfm. We propose to operate at a pressure of 10 atmospheres to decrease the velocity by a factor of ten and to reduce the pumping power.

For He the critical pressure is 2.26 atmospheres; thus, not much over-pressure is permissible. However, the vapor density at 4°K (1 atm) is about 10 per cent of the liquid density. A return pipe at 4°K and ten times the area will then not add much pressure drop and will only double the helium investment (~ 28 million ft^3 He gas STP, costing 2×10^6 and ~ 10 per cent of the annual production).

Cooling by supercooled N_2 and He^4 has been considered but was found to be much less effective. Superfluid He (He II) could be used without pumping, but only over much shorter distances.

Vacuum Pumps

The limiting pumping speed of a pipe (in cm^3/s) of diameter D and length L is [20]

$$C = 1.2 \times 10^4 D^3 L^{-1}.$$

The volume is $V = \pi/4 D^2 L$. Thus the pumping time constant is

$$t = \frac{V}{C} = 6 \times 10^{-5} L^2 D^{-1} \text{ seconds}.$$

Thus the pump capacity and pump separation depend upon how rapidly it is desired to pump out the air. For example, since $D = 15$ cm, if we choose $t = 200$ seconds (or total pump-down time of ~ 15 minutes) the pump separation must be ~ 50 m. The speed of the pumps then must be the volume of line in 50 meters per 200 seconds or 25 l/s. However, if we choose $t = 3$ hours (a total pump-down time of ~ 10 hours) the pump separation can be 500 m with a pumping speed of 5 l/s.

Appendix E

Differential Expansion

The cryogenics engineer, like his steamy brother, must reckon with thermal expansion. Ordinary metals on being cooled to 0°K contract ~ 0.3 per cent, some plastics ~ 2 per cent. Therefore, upon cooling, a long line fixed at the ends is subjected to a 0.3 per cent longitudinal stretch which, with an elastic modulus $\sim 10^{12}$ dyn/cm^2, corresponds to a tension $\sim 3,000$ atm or 50,000 psi. It is desirable to avoid such stresses in order to eliminate the massive insulating clamps and additional gear required to stretch the line and in particular to avoid the accompanying heat conduction. It is not impossible to use such techniques, particularly with low-thermal-expansion materials, but it is desirable to have the possibility of using any reasonable materials in the line. A contraction ~ 0.3 per cent on a 20-km run amounts to ~ 60 m, which is large. Fortunately, however, it is not necessary to allow for differential expansion between the parts of the structure which are at 4°K and 77°K unless very different materials are involved. (Even when the materials do have vastly different coefficients of thermal expansion, this part of the problem can be solved by the frequent insertion of bellows in the N_2 lines of, say, 0.1 per cent or ~ 50 cm every 500 meters.) Undoubtedly the simplest solution to the problem of the contraction of the cable itself is to twist the superconducting cables as they are fed into the housing. A twist which increases the length by a few per cent will allow (preferably with a compressible center) an extension of 0.5 per cent without much cost penalty. Finally, to take care of the absolute contraction of the large-diameter housing for the cables and for the helium line, bellows, ~ 1.5 m long every 500 meters, can be inserted.

References

[1] Howard, J. G. 1965. Future availability and cost of electrical energy. *Iron and Steel Engineer* September: 204.

[2] Olmsted, L. M. 1965. 14th steam-station cost survey. *Electrical World* October 18: 103.

[3] National Power Survey. 1964. *A Report by the Federal Power Commission.* Washington: U.S. Government Printing Office, p. 212.

[4] Benz, M. G. 1966. Superconducting properties of diffusion-processed niobium-tin tape. *Bull. Am. Phys. Soc. II* 11: 108.

[5] Montgomery, D. B., and Sampson, W. 1965. Measurements on niobium–tin samples in 200-kG continuous fields. *Appl. Phys. Lett.* 6: 111.

[6] Kantrowitz, A. R., and Stekly, Z. J. J. 1965. New principle for the construction of stabilized superconducting coils. *Appl. Phys. Lett.* 6: 56.

[7] National Power Survey, *op. cit.*, p. 156.

[8] Laverick, C. 1965. Progress in the development of superconducting magnets. *Cryogenics* 5: 152.

[9] Scott, R. B. 1959. *Cryogenic Engineering.* Princeton, N. J.: D. Van Nostrand Co., pp. 250–51.

[10] McFee, R. 1959. Optimum input leads for cryogenic apparatus. *Rev. Sci. Instr.* 30: 98.

[11] Spreadbury, F. G. 1962. *Electronic Rectification.* Princeton, N.J.: D. Van Nostrand Co., p. 196.

[12] *Ibid.*, pp. 177–78.

[13] National Power Survey, *op. cit.*, p. 157.

[14] Seraphim, D. P., d'Heurle, F. M., and Heller, W. R. 1962. Coherent superconducting behavior of two metals (Al–Pb) in a synthetic filamentary structure. *Appl. Phys. Lett.* 1: 93.

[15] Cline, H. E. *et al.* 1966. Superconductivity of a composite of fine niobium wires in copper. *J. Appl. Phys.* 37: 5.

[16] Bean, C. P., Fleischer, R. L., Swartz, P. S., and Hart, Jr., H. R. 1966. *J. Appl. Phys.* 37: 2218.
[17] Smithells, C. J. 1962. *Metals Reference Book*, vol. 2. Washington, D.C.: Butterworths, p. 744.
[18] *Yearbook of the American Bureau of Metal Statistics for* 1962. New York, 1963, p. 98.
[19] Bean, C. P. *et al.* 1966. A research investigation of the factors that affect the superconducting properties of materials. General Electric Research and Development Center, Schenectady, N.Y. Tech. Rept. AFML-TR-65-431, March.
[20] Scott, R. B., *op. cit.*, p. 156.
[21] Wilkinson, K. J. R. 1966. Prospect of employing conductors at low temperature in power cables and in power transformers. *Proc. IEE.* (*London*) 113: 1509.

6
Superconductors in Technology

John K. Hulm

Introduction

At the present time the highest known superconducting transition temperature is about 20° Kelvin. Since the critical field and critical currents are zero at T_c, it is usually necessary to work well below T_c in order to get useful values of these properties. While it may be possible occasionally to use liquid or solid hydrogen operating at low pressures, more commonly the best operating temperature is 4.2°K, obtained with liquid helium.

We must pay a certain price in operating at such a low temperature. For a refrigerating machine operating between 300°K and 4°K the Carnot efficiency is about 1/75, and the real efficiency is usually a factor of five lower, allowing for deviations from thermodynamically reversible conditions. This simply means that to extract one watt of power from the 4°K environment and to reject it at room temperature requires about 400 watts of power input to the refrigerator in the first place. This is a good yardstick to bear in mind when any application of superconductors is being evaluated. If one is concerned with power dissipation, for example, and the use of superconductors does not give one at least a factor of 1,000 below the room temperature power losses, the application seems doubtful. This consideration is most important in connection with the possible use of superconductors in ac power systems, which will shortly be considered.

The author is director of Cryogenics in the Research Laboratories of the Westinghouse Electric Corporation, Pittsburgh, Pennsylvania.

In spite of this rather serious thermal limitation, the dissipation of electrical energy in superconductors is frequently so low that one is willing to pay the refrigeration price. Certainly, under exactly dc conditions, supercurrents should flow without any loss whatever in the subcritical region. In a closed superconducting circuit persistent currents may maintain themselves at a constant level for an indefinite length of time. This property is a very important one from an electrical engineering viewpoint, since it permits the construction of an essentially loss-free inductor which is the proper magnetic analog of the electrical capacitor. Such an inductor is useful at high power levels for storing energy and at low power levels for storing information, that is, as a memory element.

Aside from its employment in the short-circuited condition as a storage device, the superconducting inductor is of perhaps even greater use as a direct source of flux in the form of the superconducting magnet. With the advent of Type II superconductors, this application has come to prominence during the past five years and I shall consider it in some detail.

If the inductor is combined with a superconducting capacitor to form a tuned circuit, the high frequency losses may be quite low providing that the circuit is operated in the Meissner region. Consequently it is possible to construct ultrahigh Q circuits over the range from a few hundred cycles up to the millimeter wave region and beyond. Q values of 10^5 or more are possible for microwave cavities of niobium. It may be of interest to employ such cavities in linear accelerators, where the refrigeration power is less than the power loss due to microwave losses in a normal resonator.

The switching properties of superconductors are also of technological interest. By magnetic or thermal means, a superconductor may be transformed suddenly from a zero resistance mode to one of finite resistance. This is, of course, the inverse of the mechanical switch, which, of course, transforms from a finite resistance to an infinite one. Where complete circuit interruption is required the superconductor appears to be of little use, but there are several switching applications which appear attractive. If superconducting switches are coupled with storage inductors it is possible to build a wide variety of ac-to-dc converters of fairly high efficiency. Magnetic flux can readily be transferred from one circuit to another, and the magnetic field strength can be increased or decreased at will. These devices have been generally termed "flux pumps". Their specific usefulness has yet to be demonstrated, but their characteristics appear to be so unusual as to merit a more detailed discussion.

Just over a decade ago Buck described a simple superconducting switch known as a "cryotron" for use as a bistable element in computers. A great deal of work has been carried out on cryotrons in the intervening period. By employing thin films in close proximity, the size of the device was reduced and its speed increased to the submicrosecond range. Despite these improvements I know of no computer in commercial use today which employs the cryotron. Probably this is because other room temperature devices have proved as efficacious and easier to operate.

Superconducting thin film memory elements based upon flux storage are considered to be superior to other storage devices such as magnetic thin films in the domain of very large memories, but here again I know of no commercial utilization of these at present. For this reason and also because there have been few recent developments in this area, I shall exclude computer elements from my discussion. I shall consider first the technology of superconducting magnets; second, the prospects of ac power applications of superconductors; and third, some of the interesting devices in the flux pump family.

Superconducting Magnets

The ultimate power output of any electrical machine is limited by the temperature rise of the conductors and magnetic components of the system. The conductor heating depends in turn upon two factors, the current density employed and the surface cooling. If forced cooling is employed, one can obviously work at a higher current density than in a static or purely convective situation. In the case of an electromagnet this limitation on the current density also sets a limit on the available field. It turns out to be very difficult to design a copper electromagnet with an iron core to generate much over 10 kilogauss without using some form of forced cooling.

This situation gets steadily worse with increasing field. For example, a copper electromagnet which generates 100 kilogauss in a field volume of a few cubic inches will require a power input approaching one megawatt and a coolant flow of thousands of gallons per minute.

The problem of generating high fields was greatly eased by the discovery of Type II superconductors with high critical current densities. As long as one works such a material at a current well below the critical value, and of course maintains a temperature well below T_c, there is no joule heating in the conductor and thus no forced cooling requirement such as for a copper coil.

The possibility of using Type II superconductors in such a fashion had been vaguely realized in several laboratories before 1961, but in that year some of the dramatic properties of Nb_3Sn reported by Kunzler and his associates of the Bell Laboratories gave a tremendous impetus to the field. The critical parameters of Nb_3Sn are shown in Figure 1.

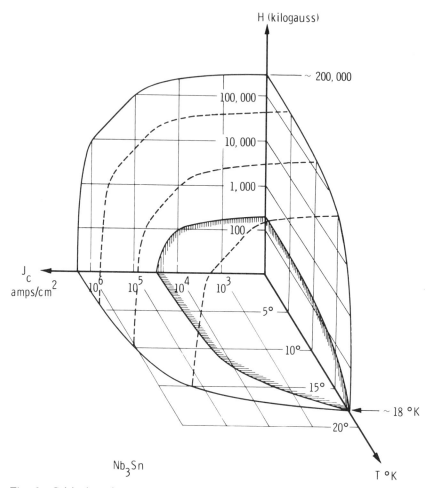

Fig. 1 Critical surfaces of magnetic field, current density and temperature for superconducting Nb_3Sn. Type I behavior occurs beneath the inner (shaded) surface and Type II behavior between the inner and outer surfaces.

A certain amount of artistic license was used in constructing this figure, but the over-all features are based upon actual data. Note that there is a change from linear to logarithmic in the H and J_c scales. The upper and lower H_c–T curves are essentially an intrinsic property of the material, whereas the J_c levels are dependent upon defects in the material and thus upon the precise method of preparation. However, it is rather hard to prepare material with J_c less than about 10^5 amperes per cm^2, which was probably fortunate in the first study of this material by the Bell workers.

The lower, shaded zone, in Figure 1 indicates the approximate position of the Meissner region where fields and currents are confined to a thin surface layer of the material. Between this zone and the upper surface is the so-called mixed state, which is thought to be punctured thoroughly by a large number of fluxoids, or small normal cores surrounded by a suitable cylindrical distribution of supercurrent. The typical spacing of fluxoids is probably a few hundred Å. It should be noted that where J_c is finite in the mixed state, the current is believed to flow internally in the material, superimposed on the local fluxoid currents. This produces a Lorentz force tending to displace the fluxoids, which is offset by energy variations due to local defects. Hence the dependence of J_c on the defect concentration.

For the design of a magnet operating in liquid helium the region of importance is the J_c–H plane of Figure 1 at 4.2°K. This is illustrated in Figure 2, together with data for two other materials which are commonly used for magnet construction. These other two materials are cubic solid solution alloys, Nb + 25% Zr and Nb + 50% Ti.

Note that the alloy conductors appear to be limited to the region below 100 kilogauss. In spite of this limitation, alloys have been employed in the vast majority of small commercial magnets constructed to date, mainly because they are ductile and can be drawn to fine wires with yield strengths of several hundred thousand psi. By contrast the compounds are brittle materials of rather low inherent strength. A few years ago this seemed to offer a serious barrier to the development of a useful compound magnet conductor, but most of these technical difficulties have now been overcome. The compound is simply deposited as a thin layer on a strong refractory metal or stainless steel tape. The tape provides the required mechanical strength while the use of only a few tenths of a mil of the brittle superconducting compound allows flexibility. Several deposition methods have been used, for example, solid state diffusion, vapor transport and flame spraying.

Having chosen a conductor or combination of conductors appropriate for the field required, the neophyte magnet designer now carried out a simple calculation such as that shown in Figure 3. Ideally one wishes to have an operating line which will intersect the critical current curve at the design field. Hence, one must choose a suitable combination of magnet length, L, number of turns, N, and effective area of conductor A to give the required slope, using an elementary formula for the field at the inside layer. Actual magnet design requires a detailed analysis of the field profile on a computer, but for the present discussion the long coil approximation is good enough.

John K. Hulm

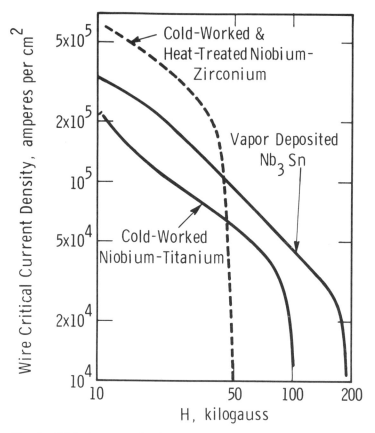

Fig. 2 Critical current density versus magnetic field for three typical superconducting magnet materials, short sample test.

When a magnet has been constructed to the design characteristics and the exciting current is increased from zero, theoretically the system moves up the operating line, starting from the origin. Eventually the boundary of the superconducting region is reached and ideally at the point of maximum field within the magnet winding the conductor starts to become normal. As Dr. Berlincourt shows in a previous chapter, high-field superconductors usually have very high normal resistivities. If this fact is combined with the high density of current in the windings plus the rather low heat capacity at low temperatures, all the ingredients are present for a very rapid temperature rise. Moreover, the joule heat generated in the normal spot tends to spill out rapidly in all directions

108

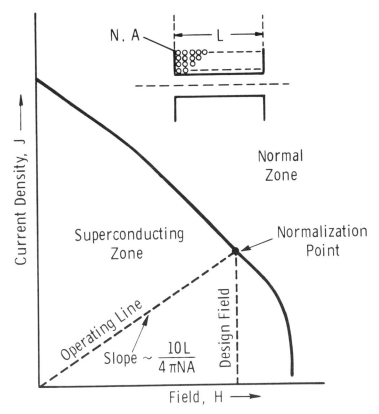

Fig. 3 Predicted operating line for a solenoid-type superconducting magnet, assuming short sample behavior.

causing a rapid growth of normality in the windings. This is illustrated qualitatively in Figure 4.

There is a well-defined velocity of propagation of the phase boundary, V, which is determined by the current density in the winding, J, the normal resistivity ρ_n, the thermal diffusivity of the winding, α_T, and the enthalpy of normalization ΔH_t. Usually α_T is highly anisotropic due to the great difference in the thermal conductivity along the conductor compared to that through the electrical insulation, so that there is a considerable anisotropy in V.

For a fully-insulated superconductor, V may run as high as 10 meters per second, so that a large piece of magnet winding may normalize in a fraction of a second. This can have several rather serious consequences.

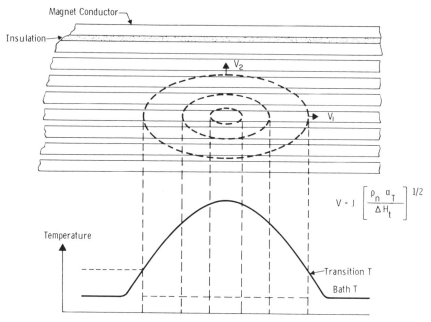

Fig. 4 Schematic representation of thermal propagation of a normal zone in the winding of a superconducting magnet.

First, the hot spot at the center of the normal zone may damage the winding. Second, the reactive impedance of the whole winding may cause a high voltage to develop across the normal zone and puncture the insulation. Last, but not least, the entire field energy is suddenly wasted as joule heat.

The first two difficulties can be reduced to negligible proportions by somewhat reducing the velocity of propagation. This can be done by simply placing a low resistivity normal conductor such as copper in intimate contact with the superconductor along its entire length. Usually quite a thin layer of plated copper is employed. When the superconductor becomes normal, the exciting current switches into the copper because of its lower resistance and thus the effective ρ_n in the velocity formula is reduced. The lower velocity of propagation results in a longer period of field decay, thus reducing the peak values of the winding temperature and voltage. The use of normal shunts is more or less standard practice in today's commercial magnets.

Shortly after the first high current density superconducting magnets were constructed it was found that the idealized design procedure outlined was inadequate due to a phenomenon variously referred to as current degradation or instability. This effect is illustrated in Figure 5. It is found that the magnet operating line fails to reach the limiting critical current curve, but premature normalization occurs at a lower current and lower field. Figure 5 serves to illustrate the fact that if the critical current density of the conductor is increased, as may be done in

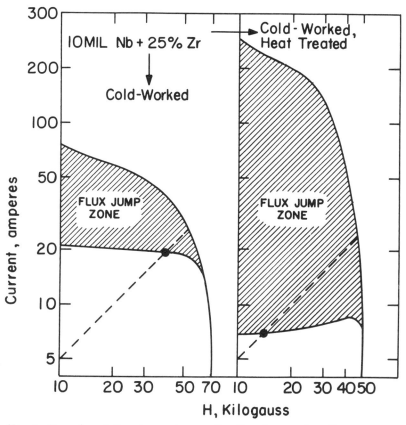

Fig. 5 Experimentally observed operating line for a solenoid-type magnet, showing zone of premature normalization due to flux jumping for two different material treatments (shaded areas).

Nb + 25% Zr by a simple heat treatment, the premature normalization effect gets worse and occurs at a still lower field for the same coil geometry.

This phenomenon is now known to be due to an effect known as "flux jumping" to which all high J_c Type II materials are prone. It was noted earlier that in these materials the current may flow through the interior. The existence of an interior circulating current produces a magnetic field gradient such as that illustrated in Figure 6 for the case of a cylinder with the field parallel to the axis. A flux jump consists simply of a rather sudden disappearance of the field gradient due to spontaneous decay of the circulating current. Frequently, as a field is swept over such a material a whole series of successive flux jumps will occur as circulating current builds up, decays and forms again.

The situation in a magnet winding is slightly more complex than that shown in Figure 6. For example, the conductor is subjected to a transverse field and hence a whole series of domains of circulating current are produced, such as is illustrated schematically in Figure 7. Experiments

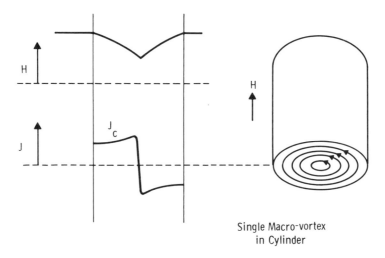

Single Macro-vortex in Cylinder

Fig. 6 Variation of the internal magnetic field and the internal current density across the cross section of a strongly pinned Type II cylinder in an axial magnetic field.

have shown that these are often elliptical in cross section due to anisotropy in the critical current density, and displaced laterally by the superimposed transport current. When decay of one such domain occurs spontaneously, the temperature rises as the current in the domain is dissipated. The resulting heat pulse tends to trigger the decay of neighboring domains and to combine with the transport current to form a nucleus for thermal propagation in the magnet winding. This is the basic cause of premature normalization in high J_c magnet systems.

It seems obvious that the specific energy released in flux jumping may be diminished by employing a small conductor dimension in a direction normal to the field. However, little progress has been made with this idea, probably because most magnets have a fairly complex field pattern in which it is difficult to preserve the correct orientation of the narrow conductor direction relative to that of the field vector.

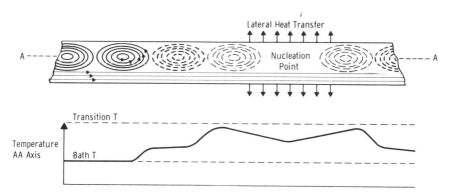

Fig. 7 Schematic representation of one instant during the nucleation of a normal zone in a magnet due to the collapse of an individual macro-vortex.

Instead of attacking the difficulty at its source, it is possible to inhibit the thermal propagation process which occurs after the jump. Stekly and Laverick have shown indirectly that the thermal propagation velocity can be reduced to zero by a combination of heavy normal shunting of the superconductor and lateral cooling of the magnet conductor.

The principle of stopping propagation can be understood by reference to Figure 8. Consider a 10-mil diameter superconductor wire surrounded by a high conductivity copper sheath and directly immersed in liquid helium. At the normalization point, 100 amperes is assumed to switch from the superconductor to the copper sheath. The power dissipation per unit surface area is plotted *versus* the ratio of copper area to superconductor area.

For heat transfer into liquid helium, up to one-half watt per square cm can be carried before nucleate boiling stops and large temperature differences occur between surface and bath. It is clear from Figure 8 that for zero copper the superconductor itself liberates about 10,000 watts per cm^2; this power is obviously going to cause rapid motion of the normal-superconducting boundary along the wire, since the lateral power loss to the helium bath is negligible by comparison.

However, as the copper-to-superconductor ratio is increased, the

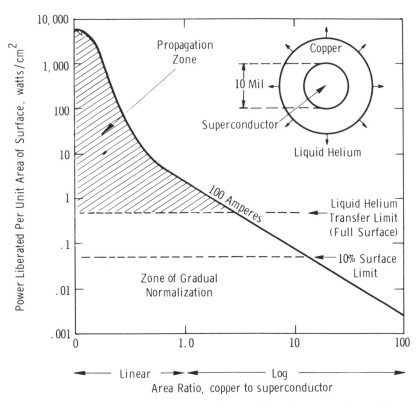

Fig. 8 Effect of switching 50 amperes from a 10-mil superconductor into a surrounding sheath of copper immersed in liquid helium. The power liberated per unit surface area of the copper is plotted *versus* the relative area of copper-to-superconductor. In the shaded area this power cannot be absorbed by the liquid helium under nucleate boiling conditions, and thermal runaway occurs.

normal resistance decreases and so does the power liberated per unit area of surface. A point is reached at which all this power can be absorbed in the liquid helium without appreciably increasing the surface temperature of the wire, and at this point propagation ceases. In actual magnet construction it is very difficult to utilize the entire conductor surface for cooling. Thus, a 10-percent surface utilization factor moves the threshold level down to the dotted line, which calls for about ten times as much copper as superconductor.

The characteristic of a magnet constructed with the necessary heavy shunting and proper cooling is that when the normalization field is reached the appearance of resistance is gradual and monotonically dependent on the exciting current. If the current is lowered below the critical value, the resistance disappears again. There is no sudden loss of field energy, a characteristic which is probably the most attractive of all for the designer of very large magnets. However, this objective is not attained entirely without cost. The heavy dilution of the winding with copper lowers the effective current density and thus increases the required thickness of the winding for a given field level. For very large magnets this appears to be a reasonable price to pay for the maintenance of field energy. It appears likely that high field–low working volume magnets will continue to be constructed for slow normalization.

A number of magnets have been already constructed on the heavy shunting-internal cooling approach, at the Argonne Laboratory and at Avco. Some very large, low-field bubble-chamber magnets are currently being planned by the A.E.C. In these systems mechanical forces begin to play a major role. One physical limit to the size is obviously reached when the hoop stress in the conductor equals the yield strength of the material. Such limits are illustrated in Figure 9 for an average current density of 5,000 amperes per cm^2, which is typical for a heavily shunted magnet, where the main material is likely to be soft copper. However, it should be noted that at a given H the critical radius is inversely proportional to J_c, and the latter may be diminished considerably below 5,000 in a large diameter magnet. Moreover, the winding could readily be given additional strength by the use of separate structural members.

In concluding this discussion of superconducting magnets, it will be remarked that the main application of these devices has so far been in research fields such as nuclear physics and solid state physics. The impact has already been a major one—for example, the number of laboratories exploring the temperature region below 1°K has increased substantially due to the availability of a relatively low cost, high field device. The utilization of these new magnets in technology is still essentially nonexistent. Perhaps the most promising area is that of magnetohydrodynamic power generation, which probably would not have been economically feasible *without* superconducting magnets. Unfortunately MHD may not be attractive even with the new magnets. Other possible areas of magnet use include homopolar generators (dc) and magnetic ore separators, but these do not seem to have progressed beyond the conceptual stage.

Alternating Current Devices

The prospects for ac application of superconductors will now be considered briefly. Most of the present power systems are constructed for 60 cps operation and it is natural to consider replacing the various components by superconducting devices. High magnetic fields are often

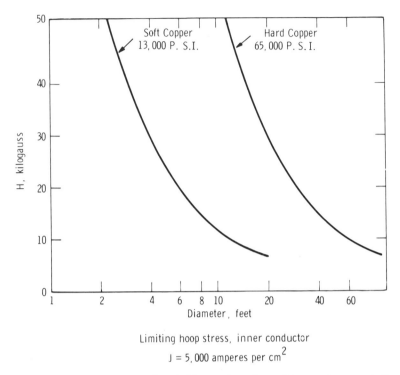

Fig. 9 Maximum operating field as a function of internal magnet diameter for a heavily shunted superconducting magnet, as determined by the maximum yield strength of either soft or hard copper. (Current density 5,000 amperes per cm².)

desired, so that Type II superconductors are important in this connection.

The behavior of Type II materials in an alternating field has been greatly elucidated by the work of Bean and London. This represents an extension of the internal circulating current concept which was referred to in connection with flux jumping. The kind of phenomena which are encountered are illustrated schematically in Figure 10, where a series of increasing field strengths is applied parallel to the axis of a cylinder of Type II material. The figure shows the expected distribution of field and current across a section of the cylinder and the resultant B–H curve.

At a low applied field H_1 the internal currents penetrate only partially into the material, in a layer which completely screens out the field from the center of the cylinder. Eventually, at a high enough field H_2, the internal current zone reaches the center and beyond this point the field

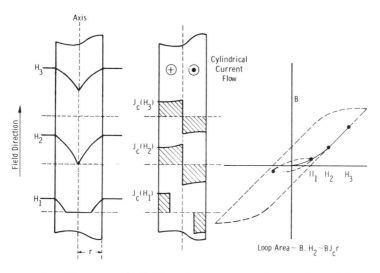

Fig. 10 The magnetic field and current density profiles within a strongly pinned Type II cylinder due to several different values of applied axial field. On the right is shown the magnetic induction resulting from a complete alternating cycle of the field.

at the axis begins to rise. The field and current pattern is substantially unchanged at a higher field H_3.

As a result of this behavior the total flux density in the cylinder lags the external field, yielding a B–H curve shown by the $H_1H_2H_3$ points on the right of Figure 10. It can readily be shown that if H is decreased from some arbitrary point above H_3, the circulating current pattern reverses itself and eventually causes B to exceed H. This results in the dotted hysteresis loop shown, which is analogous to the hysteresis loop of a ferromagnetic material. The area of the loop represents a real loss associated with microscopic eddy currents in the superconductor. If the saturation field H_2 is not exceeded, i.e., if the circulating current occurs only in a surface sheath, the hysteresis loop is smaller in area. For the saturated material the area of the loop and therefore the loss density is size-dependent.

From this type of picture one may estimate quantitatively the losses in Type II materials in ac field and currents. The detailed results will not concern us here, but a few examples will serve to indicate the general picture. Figure 11 shows some experimental data for the 60-cps loss density in four different superconductors plotted as a function of the applied current (lower scale) and the equivalent current density (upper

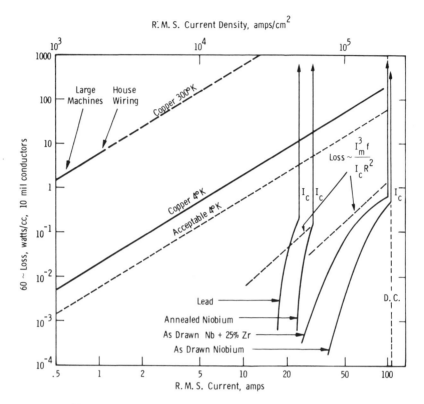

Fig. 11 Sixty-cycle alternating current watt-loss density as a function of the current in a 10-mil conductor, for copper and various superconductors at 4°K and for copper at room temperature.

scale). Also shown for comparison are the corresponding loss densities in copper wires at room temperatures and at 4°K. Temperature rise is assumed to be negligible. The loss density in the copper wires is size-independent, whereas for the superconductors it is not. This has to be continuously borne in mind.

Figure 11 indicates that the working current densities normally employed in electrical systems lie close to 1,000 amperes per cm², corresponding to a joule heating rate of about 1 watt per cc. The permissible loss is somewhat higher in house wiring which has the largest surface-to-volume ratio, and in generators, transformers, etc., would have to be set considerably lower, but for the use of forced cooling. The "acceptable 4°K" line merely shows the level of loss at 4°K which, after the refrigeration penalty has been paid, would be equivalent to the room temperature losses in copper. Clearly copper itself at liquid helium temperatures is not a good performer. However, all four of the superconductors exhibit acceptable loss densities, below the copper equivalent line, particularly if operated at half critical current or less.

The dotted curves shown just above the superconductor experimental curves are the predictions of the hysteresis loop model discussed previously. Clearly the order of magnitude agreement with theory is not bad. There is also a simple explanation for the very steep drop-off of the experimental curves. The hysteresis model of Figure 10 ignored the existence of a Meissner region at low fields where currents and fields are confined to a thin surface layer much thinner than the surface sheath of the hysteresis model. In this Meissner region the losses should be negligibly small at low frequencies. The drop-off region represents an overlap of the hysteresis zone and the Meissner zone.

The optimistic picture for alternating currents is not repeated when the application of large alternating magnetic fields to a Type II material is considered. In Figure 12 are shown some experimental data for Nb–Zr alloys up to about 1,000 gauss field amplitude. The dotted curve shows the loss density expected from the hysteresis model in the unsaturated region, and again the agreement is not bad. Data for these samples

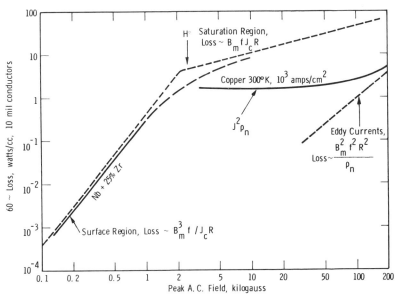

Fig. 12 Sixty-cycle alternating magnetic field watt-loss density plotted *versus* the peak field value, for Nb–Zr at 4°K and for copper at room temperature.

could not be obtained in the saturation region, owing to heating effects connected with the large loss density, but experiments on smaller samples at the General Electric Research Laboratories have demonstrated the saturation effect.

It is interesting to note that for a copper wire subjected to the same field amplitude, the eddy current losses rise according to the right-hand dotted line and clearly lie several orders of magnitude below the superconductor curve. In actual practice copper will usually be worked at a transport current density of about 1,000 amperes per cm^2, so the corresponding joule heating is included to provide a more realistic comparison. It is fairly obvious that if the acceptable superconductor losses are to be below a curve 1,000 times lower than the copper curve, the maximum alternating field amplitude must be restricted to a few hundred gauss.

The net effect of these considerations is that superconductors may be useful in quite high alternating current applications, but not in high alternating field situations. As a good example of the first category, we may cite the transmission line, where high current densities are desirable and high fields can be deliberately avoided. This point will not be elaborated in view of Dr. Garwin's paper, except to remark that both ac and dc superconducting power transmission appear economically attractive for high power levels, in excess of a few hundred MVA.

The armature of an alternator appears to be a good example of the undesirable alternating field type of application. This armature is subjected to a large alternating and simultaneously rotating magnetic field of several kilogauss. It seems mandatory to conclude that such a device will continue to be made from copper–iron.

Between these two extremes there is a whole range of situations where superconductors appear to be borderline or to present an advantage subject to overcoming a number of other massive technical difficulties. The advantage offered is almost always one of loss reduction for a device of given size. Thus for a given power level the machine can be made smaller, which is a desirable goal for space applications. Alternatively, for a given size of machine, superconductors may enable an increase of the power throughput. This is a goal of greater interest to the electrical power industry because the present maximum power levels are set by considerations such as the height of the standard railroad gauge (for transformers) and peripheral speeds or shaft whip problems (for generators).

Large ac generators in the 50 to 500 megawatt region of power output usually employ a stationary armature surrounding a rotor which, in the rotating reference frame, supplies a dc field of 15 kilogauss. Now that very large superconducting magnets appear feasible, it is possible that a high field superconducting rotor magnet can be used to increase the generator capacity. Of course the attendant rotational problems may be very severe. It is obviously easier to construct a smaller machine with a stationary dc field applied to a rotating armature. Small pilot models of this type in the size range of a few kilowatts have already been tested in several laboratories.

Power transformers can be considered in the low field category because for closely coupled windings the primary and secondary load currents are essentially 180° out of phase, and the local field at the winding can be kept fairly small. However, the power losses in a transformer are usually split about equally between the conductors and the magnetic core, so that complete elimination of the conductor losses would only reduce the power loss by a factor 2. At the present time superconducting power transformers do not seem particularly attractive.

In closing this brief discussion of possible ac applications of superconductors, it seems appropriate to remark that one of the obstacles to radical innovation in the field of large electrical apparatus is the very high degree of sophistication of the existing equipment. Over 80 years of technological development have brought these large machines to a high pitch of efficiency and reliability. They are expected to give and do give decades of trouble-free operation, often working well beyond the original design ratings. To translate the present pint-size superconducting laboratory devices into large superconducting apparatus with performance levels superior to the existing power systems will require not only great persistence but also very substantial development funds.

Flux Pumps

As was already pointed out, the lossless superconducting inductor is a versatile device. Its flux storage property permits the construction of various flux transfer devices which can be used as sources of large direct currents. Such currents may be generated either mechanically or from ac inputs. One of the main reasons for interest in these devices, which are usually termed flux pumps, is that large superconducting magnets require dc power. By the use of flux pumps it is hoped to avoid the thermal losses associated with running heavy electrical leads into the cryogenic environment from room temperature.

The general idea behind this type of device is illustrated in Figure 13. Consider the single switch (1) in which a small permanent magnet generating a field B moves across the circuit from left to right. A voltage pulse is induced as each arm of the circuit is crossed, but the load voltage pulse has a different height for the superconducting branch as compared to the normal branch. This difference is simply due to the resistance injected during passage across the superconducting arm, and a rectification effect would be produced for any magnetoresistive material in one of the arms.

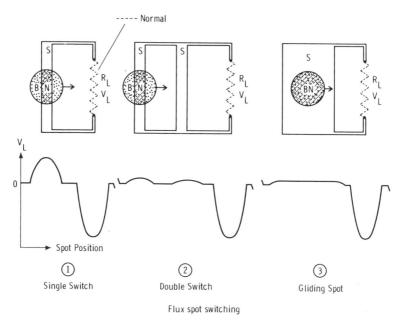

Fig. 13 The "gliding spot" principle of flux pump operation; this illustrates the asymmetry in load voltage due to passage of the flux spot, depending upon whether the circuit arm cut by the flux is superconducting or normal.

For the double switch (2), the asymmetry is even greater because the voltages induced in either superconducting arm are essentially shorted out by the other superconducting branch; there is a logical transition to a superconducting plane (3) in which the flux spot produces a normal zone surrounded by superconductive material. A gliding normal zone of this type appears to have been first used as a voltage generator by Volger and his associates at the Phillips Laboratories. The voltages for a single plate are very small and thus a lot of plates must be connected in parallel to get an appreciable output. Such a device described by Wipf is shown schematically in Figure 14. A large number of bar magnets are driven past a set of series-connected plates by a direct mechanical drive. The mechanical drive is not essential for this device, since the flux spot can be driven along by a suitably phased chain of electromagnets. Wipf has

SUPERCONDUCTORS IN TECHNOLOGY

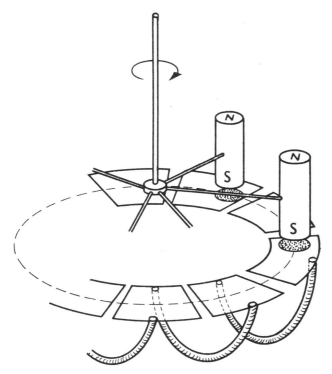

Fig. 14 A multi-segment mechanically driven gliding-spot flux pump (after Wipf).

constructed an electrically driven system based upon electromagnets connected directly to the three phase mains, as shown schematically in Figure 15.

The output of these devices depends upon the flux spot size, the speed of rotation and number of segments. Volger describes a 48-plate unit which delivered 50 millivolts into a 1 Henry inductance at 1,200 revolutions per minute. Unfortunately the output does not increase linearly with spot velocity. Volger found that at high speeds the flux spot develops a sort of "comet's tail." This may arise from either eddy current damping or Type II pinning, depending upon the type of material used for the plate. Although the flux spot generator may be quite efficient at slow speeds, its efficiency falls off rapidly with increasing speeds and this is probably the Achilles' heel of the device.

A somewhat different type of flux transfer device is possible if the flux spot or any simple source of magnetic field is used to open up a superconducting ring to admit flux from another source. This is illustrated

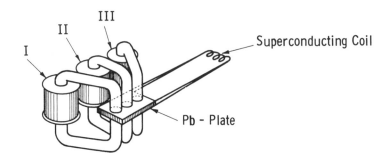

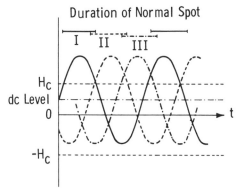

Fig. 15 A 3-phase ac electromagnetically driven gliding-spot flux pump (after Wipf).

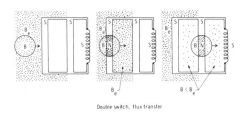

Fig. 16 Flux accumulator principle, with flux spot switching.

schematically in Figure 16, where the sequence of events is that flux is admitted to the first ring, which is then sealed off and opened up to a second circuit which may be a superconducting magnet. The reason for two circuits is that the inductance of the second one is several orders of magnitude greater than that of the first, so that flux must be pumped in gradually.

There are many varieties of the double-switch flux shifter and only two will be mentioned as typical of this family of devices. In 1938 Felici proposed the sliding contact shown in Figure 17(a). The turnstile arrangement of conductors is supposed to trap external flux and carry it around for deposition in the load. As far as is known the device was not built, but it was certainly a very ingenious idea for its era. In 1958 Olsen

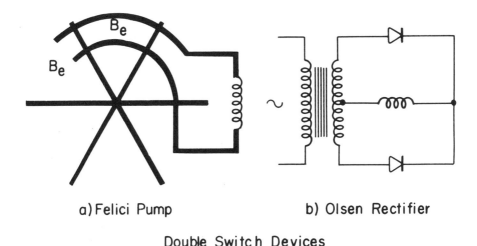

a) Felici Pump   b) Olsen Rectifier

Double Switch Devices

Fig. 17 (a) Flux accumulation by mechanical switching (after Felici); (b) superconducting rectifier (after Olsen).

invented the balanced rectifier circuit of Figure 17(b). The superconducting load is connected between the center point of the secondary of the superconducting transformer and the two superconducting rectifiers. These can be simply superconducting inductors which are driven normal by exceeding the critical current, producing a sine wave with the top chopped off. To produce rectification, a secondary coil on each inductor is employed to shift the normalization point up in one case and down in the other. Buchhold devised a variant of this circuit in which the rectifier is a power cryotron in series with a saturable reactor and a suitable control amplifier for the cryotron. Apparently this type of system is capable of rectifying with very high efficiency.

While the flux pump is an interesting laboratory device, its importance to electrical technology is not yet apparent. It appears unlikely that the efficiency will be high enough for the device to be used as a generator, but its converter possibilities, perhaps in connection with dc transmission line terminals, cannot be ignored.

In conclusion it should be noted that most of the devices discussed in this paper have emerged only in the past decade, and it is hardly surprising that most of them have, as yet, barely moved out of the physics laboratory. Before superconductivity is dismissed as a scientific curiosity, surely a greater effort should be made by electrical engineers to translate these new concepts into useful electrical machines.

7
Economic Aspects of Superconductivity
Roland W. Schmitt and W. Adair Morrison

Introduction

The preceding papers of this conference cannot fail to impress one with the dramatic developments in superconductivity during the last decade. It has been one of the most active fields of research in physics, and from it has come a series of brilliant and exciting discoveries and theories. The BCS theory, superconductive tunneling, flux quantization, Type II superconductivity, surface superconductivity, the Josephson effect and the innumerable related and subsidiary advances that have followed these seminal ones are themselves enough to explain why the field has been so attractive to scientists. And on top of this scientific interest is the possibility of important applications: cryotrons, tunnel junctions, generators, intense-field magnets, gyroscopes, transmission lines, generators, transformers, and particle accelerators are among those that have been explored. In view of these developments, we ought not to be surprised at the large effort that has been put into the field of superconductivity in the past few years.

Perhaps then it is harsh to turn the cold eye of economic analysis on this activity. All of us who have been associated with the field of superconductivity have been at such a pitch of excitement over the scientific and technical developments that we have had little time—let alone the

---

Dr. Schmitt is research and development manager of Physical Science and Engineering, and Dr. Morrison is a project analyst, Research and Development Application Service, General Electric Research and Development Center, Schenectady, New York.

inclination—to ask some of the fundamental economic questions about the field. For example, how much has been spent in the field? Or, has the cost of it been too high or too low compared with what might have been gained from alternative expenditures; has the value of the results—commercially and socially—been enough to warrant the expenditures, or is it expected to do so in the future? Only a few of those concerned with the field have asked even the simplest of economic questions, namely, how much business is going to be generated by the advances in superconductivity? How much, in saleable goods and services, is going to come from our work in superconductivity?

In some ways, this question is a source of discomfort. All of us observe a lot of money and man power being devoted to superconductivity. Of course, we have a deep belief that scientific advances of the kind that have occurred in the field are worthwhile and will ultimately pay off to society. But, it is reasonably clear that the pay-off in this field, if it is to materialize, must be in goods or services of real commercial or social value. We are uncertain about the prospects of such a return.

Certainly, the discoveries and inventions in superconductivity, though every bit as dramatic from a fundamental point of view as those in semiconductors in the late forties and early fifties, do not *obviously* hold the potential import for society that those in semiconductivity held. Indeed, the annual semiconductor business in the U.S. today is about $840 million (1965). Many of us would be quite surprised if in the next decade the business arising from superconductivity would amount to 10 per cent of that. Thus, lingering in the back of the minds of many people is a question of just how large the business in superconductivity will be, and whether it will be large enough to justify having spent so much on research and development.

The economic viewpoint has popped up elsewhere. For example, the University of Illinois in a bulletin issued by the College of Engineering in October, 1965, announced that "the University was saved three-quarters of a million dollars and considerable inconvenience by technological developments in superconductivity...." Because of the availability of high-field superconducting magnets, it had eliminated the construction of a conventional magnet room from its new Materials Research Laboratory In a burst of genuine enthusiasm the bulletin said that, "Once the phrase 'basic research' had an academic, ivory-tower ring,—Now the ring is more like that of a cash register." From the point of view of the University of Illinois, there is reason for this enthusiasm. But from the point of view of electrical equipment manufacturers who might otherwise have supplied the large, multimegawatt motor-generator sets and electrical control equipment the event may seem less than a happy one; their sales to the University are going to be less than they would have been otherwise. I dare say that if the College of Engineering had consulted the Department of Economics they might have been told of the inadequate level of demand for goods and services in our economy at that time (times have changed since that October) and they might have had second thoughts about the depressing economic effects of their $750,000 "savings." But, of course, they have probably spent the $750,000 on other items, if not on a magnet room!

There is another aspect of this question that should attract our attention. The word "application" means different things to different people in reference to a scientific discovery. To a scientist, the use of superconductivity in resonant cavities for a linear particle accelerator is an "application" of superconductivity. But to the layman, it must seem like an odd, even if correct, use of the word; he wouldn't have thought that one ought to speak of an "application" of scientific research only to other scientific research. By an application of superconductivity, he is more likely to mean its use in power transmission, in computer memories, or in guidance systems—items that have well-recognized utility in commerce, industry, defense, or in personal consumption. And yet, with more than 20 billion dollars a year being spent on research and development, we suspect that even those applications of research that are significant only for other research might nevertheless be of economic importance. This feature of the economic aspects of scientific research deserves attention in superconductivity because a good portion of the present applied output of the field finds its usefulness in other areas of research.

Finally, superconductivity, like many other areas of science today, is finding economic issues becoming more and more important. We may be dismayed that it is becoming harder and harder for a good idea and a good man—or a big machine and a good team—to get support almost automatically. But, dismayed or not, we are faced with the situation that our ideas must compete more intensely than ever before with other good ideas and with entirely different uses of the funds.

Thus, the economic issues about superconductivity are typical of those that are arising more generally in science and technology Superconductivity today is a prototype of many active, still-developing areas of research and development; it is a combination of the enthusiasm, faith, and hopes of technical people and of the uncertainty, puzzlement, and skepticism of those who supply funds.

We shall examine three aspects of this scene: first, the economic structure of the field; second, the incentives behind the financial support of the field; and third, the prospects of technological "pay-off" from the field.

Economic Structure

At the center of the economic picture of superconductivity is the group of people working on the subject. In the U.S. the members of this group are spread among institutions across the country, some spending their time in the laboratory, some at desks, some in classrooms, some in testing superconducting circuits or building superconducting coils; all have some aspect of superconductivity as the predominant subject in their working lives. As a result of their professional activity, the members of this group engender a variety of economic activities. Figure 1 depicts some of these activities.

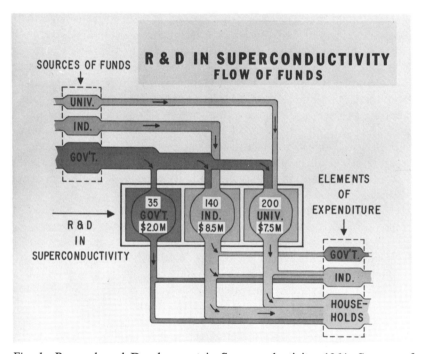

Fig. 1 Research and Development in Superconductivity, 1964: Sources of Funds; Participants in the activity—government laboratories, industrial organizations and universities—with numbers of Ph.D. level scientists and expenditure in millions of dollars in each of these segments; and elements of expenditure.

At the center of this figure is shown the institutions engaged in research and development in superconductivity. Some are government laboratories, some industrial laboratories and some universities. The figure shows that government, industry, and universities also each supply part of the funds for this work, and shows the flow of funds. Finally, the money going into research and development on superconductivity is spent on certain items, and the figure indicates these elements of cost.

Let us begin by looking at the size of this central activity, now and in the past, and at how it is growing. It is extremely difficult to get precise, or even rough, data on this subject; we have nevertheless boldly attempted to make some estimates. Since 1911, when superconductivity was discovered, we estimate that, world-wide, nearly 150 million dollars has been spent on research and development in superconductivity. Of this, about two-thirds has been spent in the U.S.

Looking at a single year, 1964, we estimate that there were about 375 Ph.D. scientists working in the field. If they were distributed among universities, industries, and government in rough proportions to the distribution of papers from these institutions, then about 35 of these were in government laboratories, 140 or so in industry, and about 200 in universities.

These numbers are shown in Figure 1.

Using these numbers, we further estimate the expenditures to be roughly $18 million, about $8.5 million in industry, about $7.5 million in universities, and about $2 million in government laboratories.

These amounts, in millions of dollars, are also shown in Figure 1.

The growth of effort in superconductivity during the first half of this decade has been spectacular. Almost three-fourths of all the papers published on superconductivity by U.S. scientists and engineers have been published since the beginning of 1961. Of the papers published in *Physical Review* on superconductivity research from 1935 to 1965, about a third of those from the universities and little more than half of those from industry have appeared in the five-year period 1961–65. Judging by the number of papers given at American Physical Society meetings, activity in universities is still increasing: from the universities, there were 40 papers in 1963; 41 in 1964; 49 in 1965; and 44 in the first third of 1966. And the ratio of expenditures incurred in 1961 and after was even larger; more than four to one.

Figure 2 shows an index of this growth: the number of abstracts of papers on superconductivity appearing in *Physics Abstracts*.

There was a slight drop in the number in 1965, but of course, an abstract appears in *Physics Abstracts* only a year or so after the paper has been submitted. To see if the drop in 1965 was confirmed by other measures and to obtain a measure of activity more nearly concurrent with the work being done, we looked at the number of papers presented

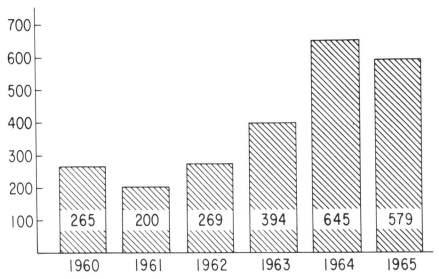

Fig. 2 Publication on superconductivity recorded in *Physics Abstracts*, 1960–65.

at meetings of the American Physical Society. These are shown in Figure 3.

There was no increase from 1963 to 1964 but 1965 does show a resumed growth. In the APS meetings of 1966, through April, 100 papers were given on superconductivity, compared with 71 for the same period in 1965, 69 in 1964, and 66 in 1963. Thus, as market analysts might say about a stock, it seems to have paused briefly but now has resumed its growth.

In terms of papers published in *Physical Review* and in *Physical Review Letters*, for 1961 to 1965, and in terms of talks given at the American Physical Society meetings of the same period, the work on superconductivity appears to be about 50 per cent in universities, a little more than 40 per cent in industry, and a little less than 10 per cent in government laboratories. These ratios are further borne out by other statistics. There were 129 attendees from 34 universities, 32 from 15 governmental organizations, and 116 from 29 industrial organizations at the International Conference on Superconductivity at Colgate University in 1963; there were 62 attendees from 24 universities, 29 from 15 governmental organizations, and 65 from 20 industrial organizations at the Conference on Type II Superconductors in 1964.

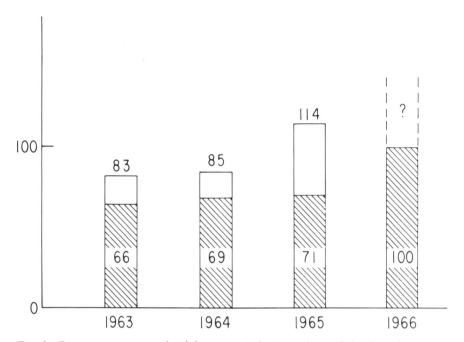

Fig. 3 Papers on superconductivity presented at meetings of the American Physical Society, 1963–66. Papers presented at meetings in the first four months of each year are shown in the shaded areas.

As you might expect, the research and development on superconductivity is relatively concentrated, a large fraction of it being done in a relatively few institutions. For example, approximately 60 per cent of all publications from industry originate in five companies (BTL, GE, RCA, Westinghouse, IBM).

Similarly, superconductivity work seems to be concentrated in 30 or 40 universities out of the 139[1] that offered work leading to the Ph.D. degree in 1964–65. At the 1963 International Conference on Superconductivity, just over 60 per cent of the 129 representatives from the 34 U.S. universities came from ten of the universities[2].

[1] AIP Directory of Physics Faculties: 1964–65.
[2] Massachusetts Institute of Technology (20); University of Maryland (10); Rutgers University, Stanford University, University of Chicago, University of California at Berkeley (7 each); University of Illinois, University of Pennsylvania, California Institute of Technology, and University of California at San Diego (6 each).

Of the $18 million we estimate as the cost of research and development in superconductivity in 1964, about $6 million went for the salaries of the principal scientists and engineers leading the work. But there are, of course, a host of other people whose work is devoted to these projects: assistants, technicians, machinists, glassblowers, all apply their efforts directly to the needs of the projects. In addition, there are secretaries, administrators, accountants, janitors, plumbers, etc.—the so-called overhead—whose services are needed to keep the research and development going even though they do not contribute directly to the projects. Unfortunately, it is difficult to make an estimate of these elements of cost. We have, without any great confidence, estimated the total salaries and wages at about $12.5 million. However, in our examination of the economic consequences of research and development on superconductivity, we should note that the salaries and wages paid to individuals because of their contributions to this activity do in turn enter the economic stream of consumer goods and services that the individuals purchase. Besides these personal expenditures engendered by superconductivity, there are, of course, also purchases of instruments, liquefied gases, pencils and paper, tools, power and water, airline tickets, etc. that also result from the research and development on superconductivity; these expenditures enter the economic stream of industrial and commercial activity. Finally, of course, some of the expenditures of industry on superconductivity are for taxes on the grounds and buildings in which the work is done, and these enter the economic stream of governmental activities. The only point we wish to make in remarking on these rather obvious elements of cost is that this activity, like any activity supported by our economy—like professional football or opera or trips to the moon—translates that support into further economic demands for goods and services. In theory then, the activity on superconductivity is a pump in the economic stream, albeit a rather small one compared to, say, the moon race.

We might close this discussion of the economic structure with some estimates, admittedly inexact, but, we hope of the right order of magnitude, of where the $18 million, applied to superconductivity in 1964, came from and where it was spent. These estimates are shown in Table 1.

Government Incentives

We have estimated that around 60 per cent of the $18 million expenditure in 1964, or approximately $11 million, was supplied by the federal government. About $2.0 million of this was used in its own laboratories and the remaining $9.0 million distributed between industry (about $4.0 million) and universities (about $5.0 million). Why is the

TABLE 1

Estimates of Expenditures for Superconductivity Research and Development, 1964
(Millions of Dollars)

Sources	Performers			Totals
	Government	Industry	University	
Government	2.0	4.0	5.0	11.0
Industry	—	4.5	—	4.5
Universities	—	—	2.5	2.5
Totals	2.0	8.5	7.5	18.0

government spending such a large amount of money on superconductivity? As far as we know, no superconducting device is used in a weapon in Viet Nam, nor is any used in the war on poverty, nor in the space program. Superconductivity at present apparently provides no *material* benefit for any governmental activity, save for research itself. Clearly then, if the expenditures are justifiable, they must be for intangible benefits or for future benefits—for *expectations*. And are the expenditures by government at approximately the right level?

Let us admit at once that a large part of the short range future benefits will continue to be in other government-supported research and development programs: high-energy physics, plasma physics, space research. We do not wish to examine the economic basis for those programs in the present paper. Let us assume that a valid economic and political decision has been made to support these areas and that value is given to research and development in superconductivity because of its usefulness in these other programs.

Some of these areas are shown in Table 2.

The first few items listed are ones in which the pay-off of successful research and development in superconductivity would be indirect in that it would depend on the pay-off of the other research programs. The economic complexion of these programs makes it exceedingly difficult to attach a dollar value to the successful development of superconductive devices for the programs. Nevertheless, some of the federal dollars spent on superconductivity in each of the three sectors—government, industry, and universities—can be attributed to the incentive of these programs.

TABLE 2

SUPERCONDUCTIVITY

Prospective Results of Value to Government

For Federally-Sponsored Research (Indirect Pay-Off)

High Energy Physics
    Resonant cavities of linear accelerators
    Beam-bending and focusing magnets
    Bubble and spark chamber magnets

Plasma Physics
    Plasma containment magnets

Space Research
    Magnetic shielding
    Magnets for ion—or plasma—propulsion devices
    Gyroscopes

Solid-State Physics
    High-field magnets

We should, perhaps make one remark concerning federal research and development expenditures in industry on superconductivity. A reasonable estimate of the potential sales by industry of superconducting materials and coils for federally sponsored research programs does not, *by itself*, make a sufficiently attractive business picture, considering the uncertainty and risks involved, to warrant the investment by industry of the total amount required for research and development. This fact is the underlying reason why it is prudent for the government to support some of the research and development that industry must do if it is to develop and supply the materials and coils needed. This situation, of course, is quite different from conventional business areas—such as power generation, transmission and distribution equipment, or polymers, or appliances, or pharmaceuticals—in which the price one can get in the market for new or improved goods is usually enough to pay for the research and development needed to provide these goods.

The situation is more nearly like that in the market for aircraft, electronics systems, or weapons in which, too, research and development is supported as a separate activity, not as part of the market price of the product.

Governmental interest in superconductivity extends beyond the results that might be of value to other research programs. There are also areas of direct pay-off. Some of these are indicated in Table 3.

TABLE 3

SUPERCONDUCTIVITY

Prospective Results of Value to Government

For Other Federal Interests (Direct Pay-Off)

Power Transmission
  Large blocks of power by cable

Computers
  Large computer memories

General Welfare of Science and Technology

Education
  Graduate training
  Teacher stimulation

The possibility of transmitting large blocks of power by superconducting cable is of interest to the government as well as to private industry. Superconducting cables, should they become feasible, might help prevent our landscape from being laced with power transmission lines. While it is unlikely that either *cost* or *reliability* of electrical transmission, by themselves, would be enough of an incentive to use superconducting cables, we are beginning to see higher values placed on the appearance of our landscape. Who knows how much we might be willing to pay for scenic beauty as an expanding population and an expanding technology put heavier pressures on it!

Another governmental interest in superconducting power transmission might be in the problem of conveying large blocks of power to the center of large urban areas. As the demand for power increases and the access routes become more precious there will be a growing economic incentive to find new ways of conveying large amounts of power into metropolitan areas. We list this as an area of governmental interest since the whole question of urbanization and the organization of metropolitan areas is, of course, a prime area of governmental concern. However, in this case, economic forces alone, without any further major social or political decisions, might provide the incentive to devise new technological solutions to the problem, whether or not they involve superconductivity.

Computers are, of course, a potential area of application of superconductivity that is important to the government. The possibility of building novel memories using continuous superconducting films or cryotron arrays is an area of interest that the government shares with other sectors of the economy.

Beyond the rather specific reasons for governmental interest in the field of superconductivity, there are also more general ones. It must have—as it does—an intense concern for the general advance of science and technology. Let us look at this concern in a bit more detail, especially with respect to the general welfare of basic science.

In recent years a good deal of political turmoil has arisen over the allocation of research support. Unfortunately, the signs indicate that we are reaching a limit to the relative amount of money that the country is willing to put into science. The response of the scientific community to these pressures has been uniform: we have pointed again and again at the unchallengeable fact that basic science has been a remarkably fruitful source of material benefits. If we haven't always noted that technological and organizational genius has shared in producing these material benefits, we may perhaps be excused for the oversight. In the U.S. at least, basic science has in the past been the junior member in this productive partnership and may deserve a larger share of attention. Unfortunately, however, it appears that the general argument for the value of basic science is becoming little more than a liturgy which, even though true, is not convincing except to those who are already believers. It is becoming increasingly clear that the scientific community is proposing more interesting and worthwhile projects, problems, questions, and fields to work in than our society is willing to pay for. Even if all of the proposed research would yield material benefits, society might still not deem this sufficient to warrant supporting all of the proposals; alternative expenditures might seem to promise even greater benefits.

How, then, is the government to handle its role in promoting the general welfare of science in the face of this competition of worthwhile ideas and, of special concern to us today, where will the subject of superconductivity find itself in this competition? We believe that the scientific community will be required, increasingly, to make the case for support of specific areas of science and technology instead of relying on the general sermon that "science is good". We are happy to observe that this is being done. A variety of reports, exemplified by those being issued by the committee on Science and Public Policy of the National Academy of Sciences, is beginning to appear in which a field or a program area is examined in depth. These reports give substance to the alleged benefits of the particular sciences and provide a more sound basis for choice by those who must choose. With respect to superconductivity, this symposium sponsored by the Division of Physical Sciences of the University of Chicago is a superb means of bringing together the kind of information needed about a particular subject—from the most fundamental to the most practical. It is particularly gratifying to see this broad spectrum of interests promoted by a university.

Now let us turn to the role of the government in supporting research at educational institutions. The expenditures at universities are frequently ostensibly for research, not teaching. However, much of the money is granted to individuals because they are good research scientists who are also good teachers. In graduate training, research and teaching are inseparable, and it does not matter much what the subject may be as long as it is intellectually challenging. Thus, part of the five million dollars spent by the government on research and development on superconductivity in universities is not really spent because the subject is superconductivity but rather because that happens to be the subject of interest to the professor being supported. In essence, support is being given to people engaged in superconductive research and development— to some members of the group we are considering as our prime economic entity—because they spend part of their time in another activity already recognized as economically valuable, i.e., in teaching graduate and undergraduate students.

Also, some of the expenditures on research and development in universities should be attributed solely to the purpose of teacher stimulation, not to the expectation of substantive scientific results from the expenditure. On the whole, the stimulation that a teacher provides to his students, the insight he gives them to the subject and the enthusiasm he imparts to them for the pursuit of knowledge is immeasurably increased by his own activities in research. Therefore, this research, undertaken as a stimulus to teaching, should not be subjected to the same standards of value as research that is undertaken because of the expectation of direct material benefits to society.

Industrial Incentives

Turning now to industrial expenditures on superconductivity, we estimate them at around $4.5 million in 1964, and this money was probably spent almost entirely in industry's own laboratories. In addition, industry accepted and spent in its laboratories about $4 million of government money. What is the incentive to industry to spend this amount?

First of all, high-field superconducting materials and coils have become a business. In 1964, the sale of wire and coils to final users probably amounted to around $1\frac{1}{4}$ million and in 1965 to about $1\frac{1}{2}$ million. If a business this size returned as much as 10 per cent after taxes on sales and if we allowed it a very generous 20 to 1 price-to-earnings ratio, it would be worth a total of only about $3 million. Thus the present and prospective sales of superconducting coils, barring further developments in the use of intense magnetic fields, are inadequate to explain the current level of expenditures by industry. What, then, does explain them?

Some of the expenditures are undoubtedly defensive; they are spent because the developments in superconductivity threaten conventional businesses. Industry must therefore learn enough about the field either to be assured that the threat is only fancied, or to ensure that it will be able to provide the new products that will replace or complement its old ones. Thus, at least a small portion of the expenditures is justified as a kind of insurance on present businesses. Some of the money spent on superconductive memories, on high-field coils, on superconductive transformers can be attributed to this incentive.

Let us dwell on this theme a bit longer as illustrated by the economics of intense magnetic fields. The advent of high-field superconducting materials has caused a revolution in the technology of intense magnetic fields. At this point, we do not wish to inquire into the reasons why such fields have value. Let us accept that fact for the moment. Prior to the discovery of high-field superconductors such fields cost quite a lot of money and there were only a few places[3] in this country that had facilities to produce them. However, high-field superconducting materials have reduced the capital cost of intense magnetic fields by a factor of 10 to 100. Part of this cost reduction has come from eliminating the need for expensive motor-generator sets that provided the power for conventionally-produced intense fields. The manufacturers of these motor-generator sets are therefore faced with the loss of income from the sale of such equipment. The introductory paragraphs gave the example of the university that had saved three-quarters of a million dollars on a building to house conventional equipment for intense fields. Manufacturers of conventional equipment retrieve part of the lost sales if they enter the business of supplying the new equipment but they still lose practically all of the income on each unit if the price has dropped by a factor of 10 to 100. Their income will equal or exceed the prior income from the sale of conventional equipment only if the market grows by more than enough to offset the drop in price; i.e., if there is a

---

[3] Eight laboratories could produce steady fields of a 100 kG, or more. (National Magnet Laboratory; BTL; University of Pennsylvania; Naval Research Laboratory; NASA; Lewis Research Center; Jet Propulsion Laboratory; University of California at Berkeley; and Research Center of the Southwest.)

highly elastic demand for intense magnetic fields. It is difficult to tell if this is what has happened with intense magnetic fields. But certainly there has been some elasticity shown in the market, and intense magnetic fields are much more widely used than before the advent of high-field superconductivity.

High-field superconductivity illustrates another important point about technological advances that is frequently overlooked. One often hears that many of our current commercial or social problems are not really technical but economic. For example, this has been said about some aspects of air pollution, urban transportation, etc. However, this might also have been said about intense magnetic fields prior to the discovery of high-field superconductivity. There was no technical barrier to the production of intense fields. Their production was quite feasible even if quite expensive. One might just as carelessly have said then that the production of such fields was not a technical problem at all but simply an economic one. But, of course, the story of high-field superconductivity dramatically illustrates that the economic problem of cheap fields was solved by a technical breakthrough. To have said that the problem was purely economic would have been grossly misleading. Until a problem has been solved from the economic point of view it must continue to be regarded as a technical problem as well. Science, technology, and economics have become so closely allied with one another that it can hardly be said that problems in one domain exist without accompanying problems in the other domains.

Consider now the remaining incentives to industry to support research and development on superconductivity. A number of potential uses of intense magnetic fields that could become quite important have, of course, already been identified. Some of these uses are indicated in Table 4.

TABLE 4
SUPERCONDUCTIVITY
Prospective Results of Value to Industry

USES
Power transmission lines
Transformers
Motors and generators
Computers
MHD generators
Fusion generators

For example, should either fusion power generation or magnetohydrodynamic power generation become feasible it is certain that they would require superconducting coils. Industry is interested in these potential applications. However, it is clear that the industry that is investing in superconductivity for those reasons is also betting a little bit on new inventions or discoveries turning up that will lead to even wider use of intense fields.

Another area of potential application is the use of superconductivity in electrical equipment: motors, generators, transformers, transmission lines, etc. Many of these applications have been considered earlier in this conference. There are formidable problems in, and barriers to, the adaptation of superconductivity for these uses. If the research and development required to make superconductivity feasible in these applications should prove to be too great, then the expense of this research and development will be limiting. Technology may indeed stop evolving along a given line if the barriers to continued evolution are too great. We do not yet know if in electrical equipment superconductivity will prove to be feasible but too expensive to develop.

Another area of potential application of considerable interest to industry is to electronics. In spite of the problems that have been encountered in the development of cryotron and continuous film memories, there are still many laboratories both here and abroad devoting effort to their development. The potential pay-off of these is great enough to provide continued incentive to work in this area. And then, too, the potential uses of tunnel junctions, particularly operating in the Josephson mode, are beginning to be explored, and there are enough exciting possibilities to warrant continued expenditures here.

One factor that dictates continued expenditures in the field of superconductivity, although it alone would probably warrant a smaller expenditure than the current one by industry, is that almost no one believes that all of the technologically important inventions and discoveries have yet been made. Is $18°$Kelvin the upper limit on transition temperatures of superconductors? Are there feasible ways of reducing or circumventing the ac losses in high-field materials? These are intriguing questions, still unanswerable with certainty, that keep many of those supplying the money as well as those doing the research intrigued and allured by the field.

University Incentives

We have estimated university expenditures on superconductivity research and development—largely on research—at about $7.5 million in 1964. Of this, about $5 million came from the government in a variety of ways, and about $2.5 million was provided by the universities, mainly in the university-paid time of professors and other permanent staff.

Superconductivity research has gone on in a few universities ever since Kamerlingh Onnes' discovery in 1911. This research received a great stimulus with the introduction, about 1948, of the Collins–McMahon helium liquefier, which made liquid helium easily available to many more institutions than before. (There were about three times as many papers on superconductivity in *Physical Review* in the years 1949–51 as in the years 1946–48.) The introduction of a superconductive switching element, the cryotron, by D. A. Buck, about 1955; the development of the BCS theory, about 1957; the discovery of superconducting tunneling, by Giaever, in 1960; and the verification of the high-field properties of niobium–tin, by Kunzler and coworkers in 1960: all these have stimulated, over the past ten years, a large and growing amount of research in the universities.

The commercial availability of coils to reach fields of 50 to 60 kilogauss from 1963 on, and of coils to reach fields of 100 kilogauss or more this year, has aided research in solid state physics in general, and research in superconductivity in particular.

The Josephson effects, predicted by a previous speaker at this conference, have provided further stimuli to research activity in the universities.

We have estimated that there were, in 1964, about 200 of the professorial staff of the universities engaged in research on superconductivity. This is a guess—based on knowledge of some of the activities of a few universities, and on extrapolation from meeting attendance. The figure may be a little high or a little low. If all of the 200 were physicists, the number would represent a little more than 2 per cent of the physicists reported by the AIP as engaged, in 1964, in colleges and universities in research and development (9,368); or would represent about 4 per cent of those physicists in university jobs that have a Ph.D. (4,793: 2,279 in research and development, 455 in administration, 2,059 in teaching). However, as well as physicists, there are in universities a few other men in metallurgy, and, possibly, a few in chemistry and engineering, who are conducting research and development activities in superconductivity. And, in addition, of course, there are many graduate students working in this field; they were not included in the above estimate, but help to account for, if indeed such accounting is necessary, the numbers of papers that are appearing, and the amounts of money that have been estimated as spent in universities.

Decisions to initiate, increase, or maintain superconductivity in a university are perhaps somewhat easier to make than in an industrial or governmental organization. The decisions are individual rather than organizational, and the "pay-off" is immediate, certain, and appropriate. The tasks of the university, to add to the body of scientific knowledge, and to give graduate students experience in research, are both accomplished through activities in superconductivity research. And the development and use of superconductivity phenomena as tools for research that is valuable and useful to the universities in their principal tasks. For example, the invention of development of the superconducting cavity linear accelerator, at Stanford, is a notable achievement that provides another potentially useful tool for high-energy physics research.

As an aside, we might remark that the burgeoning and exciting growth of superconductivity research has greatly aided the research professors in their confrontation of a major problem alluded to by A. B. Pippard in his talk to the Superconductivity Conference at IBM in 1961: it has provided them with a series of worthwhile research problems for their first-rate graduate students.

To summarize briefly, and to lead into our next topic: we have had, and have, a considerable number of people active in superconductivity research and development; we have spent, and are spending, considerable sums of money provided by the government, by industry, and by the universities; we have had a substantial growth in our scientific knowledge and understanding; we have had less success, so far, in our practical applications, but we have great expectations. Let us now look at these expectations and their current status.

Status of Expectations

From the amorphous cloud of expectations, a few technological applications of superconductivity are beginning to crystallize into more or less visible form.

1. *Magnets*

(a) Small-bore magnets, for research using high magnetic fields, with bores up to a few inches in diameter, and with central fields of up to 130 kilogauss have been built and operated.

Magnets with fields of up to 60 kilogauss have been available commercially for several years, and in the past year, magnets with fields up to 80 and 100 kilogauss, with field homogeneities of 1 per cent or better, over volumes of an inch or more in diameter, have been offered for sale. Higher fields, larger bores, and greater homogeneities are in prospect. Such magnets offer fields three or four times greater than have been available in conventional laboratory electromagnets, and allow many research groups to have, at relatively modest cost (~$35,000–$50,000), high-field facilities previously only available at 8 or 9 centers in the country.

(b) Magnets for bubble chambers, from a few inches to about 14 feet in internal diameter, to give central fields of up to 60 or 70 kG in smaller, and up to 20 to 40 kG in larger, bores are planned.

A magnet with 10-inch bore and 43 kG field has been built, and, with an insert section, it has produced 67 kG in a 7-inch bore. Magnets of 20 inches and of a meter in bore are being designed and will be under

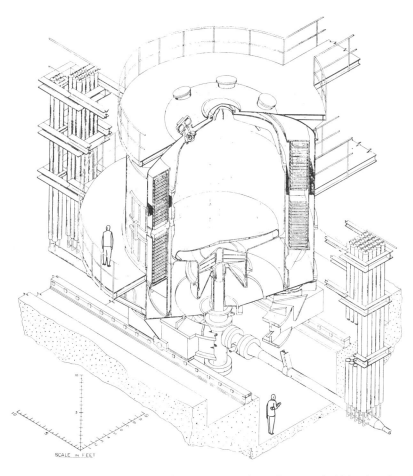

Fig. 4 Artist's conception of proposed 14-foot-diameter bubble chamber for Brookhaven National Laboratory.

construction shortly; magnets of up to four meters in bore, for very large bubble chambers, are contemplated. A very large bubble chamber magnet, now under consideration at Brookhaven, is shown in Figure 4. One of similar size is under consideration at Argonne National Laboratory.

Success with the smaller magnets will be a strong stimulus to the construction of the larger ones. The power savings, the higher fields available, and the simplification in construction and use through the removal of iron make such magnets very attractive.

(c) Magnets for accelerators. Superconducting magnets for beam-focusing and for beam-bending in large accelerators are under intensive study. A small quadrupole magnet has been built; its performance and the problems of design, cost, and operation in scaling it up to the required size are being studied. The feasibility and economics for beam-bending magnets are also being considered. For large accelerators, great savings in power, size, and weight, and, possibly, in space required for the accelerator may be possible.

(d) Magnets for MHD power generators. At least one large magnet, $\sim$ 5 feet long by 1 foot in bore has been built, and a larger one is planned.

It has been said that without superconducting magnets, MHD power generation is impracticable; it may not be practicable, in this country, with them; but those interested, here and abroad, will probably continue with the development of the magnets.

(e) Magnets for fusion research. A superconducting "baseball-seam" magnet, 10 inches in internal diameter, has been built at Livermore; one with an internal diameter of a meter is to be made. A rather large "magnetic bottle", using two 150-kG, 6-inch-bore magnets, with eight 20-inch diameter coils between these, is being built for NASA's Lewis Research Center. Although for many fusion experiments, the requirement of rapidly changing magnetic fields cannot yet be met with superconducting magnets, there are a few experiments in which they can now provide larger, and better-shaped fields, at considerably less power input, than can copper-wound magnets.

(f) Other magnets. A 6-foot diameter superconducting coil, to provide a modest magnetic field, possibly useful in shielding space vehicles from charged particles, has been built and considerably larger structures for this purpose have been studied.

Small superconducting magnets have been used in masers. Larger magnets are being considered for balloon-borne cosmic ray experiments.

In summary, the largest volume usage in terms of dollars, is at present, in the research magnets of small bore; this may change as the magnets under development and under consideration in, and for, the AEC become successful.

2. *Superconducting Linear Accelerator*

A 4-inch superconducting linear accelerator section has been operated successfully at Stanford. This promises electron accelerators of higher duty cycle (by a factor of 1,000), and of greater energy per unit length than have, up to now, been possible.

3. *Flux Pumps*

As a means of providing very large currents (up to 1,000 or 10,000 amperes, or more) for superconducting magnets, without having to bring such large currents through heavy leads into the liquid helium bath in which the magnets are immersed—with consequent heat leakage and helium loss—several types of flux pumps have been developed. These convert a small input current (an ampere, or so) into a large circulating current. The GE version uses cryotron rectifiers, requires an input current of 1 ampere, is about the size of a grapefruit, and can produce more than 1,000 amperes of circulating current. Pumping action can be interrupted, leaving a persistent current flowing. The flux pump can also be reversed to reduce the current in the magnet to a lower level, or to zero. Several models have been tested, very successfully; commercial models will soon be marketed.

4. *Superconducting Memories*

Two types of superconducting memories have been under development for some time: memories based on cryotrons and memories based on a continuous film.

Although development work on superconductive memories is still continuing in some companies in this country and in Europe, the difficulties in fabrication are formidable, and the feasibility and utility of these computers is still uncertain. Developments in the programming and the economics of current computers may, in effect, design around the potential advantages of cryogenic memories before the development can be the basis for commercial use.

5. *Gyroscopes*

Gyroscopes based on the use of a superconductor (niobium) have been under development for several years. Although potential advantages are foreseen, they have not yet been fully realized; in the meantime, other types of gyro, on which greater development effort has been concentrated, are rapidly approaching the performance foreseen for the superconducting gyro.

6. Bolometers

Superconductive bolometers have been developed. They have found little application, and they do not appear to be technologically or economically very important.

7. Transformers

Although at least one transformer using superconducting windings has been built (15 kw, McPhee, ADL ~ 1961), these are still in the study stage. The one built was small, 15 kw; the construction used lead wire conductors with interleaved primary and secondary windings, was expensive; the results may not have been very satisfactory since further development was not done. The feasibility of such larger transformers, using Type II superconductors, is still under study: it appears that overall losses (including power for refrigeration) are about the same (within a factor of 2) of conventional transformers; the size may be considerably reduced but the cost of materials may be higher by a factor of 10. There are severe problems in thermal insulation of the windings, in bringing in power leads without excessive thermal leakage, in protection against excessive load—as from a shorted line, and in providing highly reliable, unattended operation of liquefiers for the cryogenic fluid. Except for the possibility of building units of larger rating than can now be *shipped*, little advantage is apparent.

Based on what we know now, superconducting transformers do not appear to be a likely application of superconductivity in the near future.

8. Transmission Lines

Superconducting transmission lines for power have been under discussion for several years, and are still being studied. The design and economics of one such line is the subject of a previous paper at this conference.

Such transmission cables do not appear imminent; most interest now is in normal cryogenic cables—for which the problems are less difficult, though the economic advantage is perhaps no better defined or understood.

9. Generators and Motors

Experimental superconducting generators of small size have been built and a 50-kw generator is being developed. For commercial power generation, interest lies in generators with capabilities of 500 Mw, or more, and, principally, in the possibility of generators of physical size no larger than the largest that can now be shipped, but with greater generating capacity. The problems of building such generators are now being studied.

Thermal insulation of the conductors is, of course, a major problem. Combined with this, however, is the problem of restraining the conductors, in their insulating sheath, in a fixed position against the severe mechanical and magnetic forces acting on them, without, at the same time, degrading the thermal insulation. Because Type II superconductors have large ac losses, and Type I superconductors go normal at the magnetic fields required, only the dc part of the generator—the field coils—appears amenable to the use of superconductors. In large machines, the field is wound on the rotating armature. The use of superconductors on a rotating armature increases the problems of mechanical restraint and of thermal insulation, makes more difficult the problem of bringing large currents from slip rings at room temperature into a cryogenic environment ($\sim 4.2°K$ to $15°K$) without excessive heat transfer, and adds the problem of maintaining a cryogenic environment in the windings of a rotating armature, e.g., bringing in liquid helium or hydrogen from the liquefier, and removing liquid or gas to the liquefier for recooling to the appropriate temperature. Variation of current in the coils is necessary, but the rate must never be such as to cause the superconductor to go normal; this poses a nice control problem.

Although the problems would be less for the superconducting portion if the dc field were placed in the stator, the problem of removing ac power from a rotating armature by means of slip rings is one to which designers have not found a solution in large generators.

10. *Superconducting Lenses for Electron Microscopes.*

Some work is being done on using superconducting material to form magnetic lenses of high stability. Considerable development is still required; the improvement might be valuable and the scientific impact of success might be considerable; the amount of material used would be small.

Conclusion

In conclusion we should return to some of the questions asked in the Introduction. Has the amount spent on superconductivity been too high or too low compared with what might have been gained from alternative uses of the money? Has the value of the results—commercially and socially—been enough to warrant the expenditures, or are they expected to do so in the future? These were, perhaps, the most important questions posed in the Introduction and they were the least satisfactorily answered in subsequent sections. They are in reality unanswerable without an equally detailed examination of the many other fields of research and development. However, our intuitive conclusion, after this examination, is that superconductivity would certainly be neither the first victim nor the last beneficiary of a marked change—down or up—in the level of expenditures on research and development in this country.

The rate of appearance of new ideas in superconductivity has been high, and its stimulus and service to other areas of science has been notable. It has also been an exciting and interesting vehicle for training students in scientific research. These results are ones of real value to our society and we dare say that sizeable expenditures in other fields have brought fewer returns of this kind than those in superconductivity.

The material and commercial benefits of superconductivity are only just beginning. The use of high-field superconductivity for intense magnetic fields seems firmly established, but it is not yet clear just how intense fields will become useful beyond the uses in research and development itself. The future of other applications of superconductivity is faced with large risks and uncertainties. Of course, most radically new technologies look risky, uncertain, and unpromising during the course of their emergence. It is a characteristic phase in the life of a new technology, but it is characteristic of those that have only a minor impact on society as well as those that succeed dramatically. The outcome for superconductivity is not yet clear, but that minor success is certain and dramatic success still possible cannot be denied. That fact is itself enough to warrant a further fraction of the support of the field.

Thus, we conclude, imprecisely but more confidently than before this review, that the sums allocated to superconductivity in comparison with other fields of research and development have been about right.

Acknowledgements

We are greatly indebted to many colleagues at the General Electric Research and Development Center for assistance and information, and, especially, to C. P. Bean, Paul S. Swartz, and Mrs. Ethel Fontanella.

Appendix 1

Estimate of Worldwide Expenditures on Superconductivity

1. Worldwide publications, excluding USSR, Great Britain, Australia, Canada, and USA
   1911–65, at least 1,550 papers, at an estimated average expenditure of $10,000/paper $15 million

2. Great Britain and Australia
   1928–65, at least 800 papers, at an estimated average expenditure of $15,000/paper   12 million

3. USSR
   1940–65, at least 1,300 papers, at an estimated average expenditure of $16,000/paper   20 million

4. Canada
   1924–65, at least 65 papers, at an estimated average expenditure of $16,000/paper   1 million

5. USA
   1921–60, at least 800 papers at an estimated average expenditure of $25,000/paper   20 million
   1961–65, at least 2,600 papers at an estimated average expenditure of $30,000/paper   78 million

                                     Total   146 million

Numbers of papers are based on bibliography compiled by Mrs. E. Fontanella, of this laboratory. Dollar estimates of cost per paper are on 1966 dollars, and are very rough—a little better, perhaps, than guesses. They are intended to include both direct and indirect expenditures in doing the research, and to take some account of research and, particularly, development that does not culminate in publication. Readers may, of course, substitute their own estimates.

Appendix 2

Publications in the USA on Superconductivity and Attendance and Publications at Two Conferences

A. *Physics Abstracts*

Number of papers on superconductivity listed in the index of *Physics Abstracts*, for each year, for the years 1961–65.

Year	No. of Papers
1961	200
1962	269
1963	394
1964	645
1965	579

B. *Physical Review*

Numbers of publications on superconductivity in various periods.

		Source			
Period	Universities	Industry	Gov't	Foreign	Total
1935–60	329	85	78	52	544
61	31	14	2	3	50
62	23	27	6	6	62
63	19	26	5	3	53
64	42	16	2	2	62
65	46	27	4	7	84
1961–65	161	110	19	21	311
1935–65	490	195	97	73	855
					(USA 782)

C. *Physical Review Letters*

Number of publications on superconductivity.

		Source			
Period	Universities	Industry	Gov't	Foreign	Total
1961–65	101	85	15	22	223
					(USA 201)

D. *American Physical Society Bulletin* (Meeting Programs)
Papers given on superconductivity at meetings.

		Source		
Year	Universities	Industry*	Gov't	Total
1963	40	38 (32)	5	83
1964	41	39 (22)	5	85
1965	49	56 (24)	9	114
1966 (4 mo)	44	41 (29)	15	100
	174	174 (107)	34	382
Year (1st 4 mo)				
1963	32	30 (22)	4	66
1964	32	35 (21)	2	69
1965	28	34 (25)	9	71
1966	44	41 (29)	15	100
	136	140 (97)	30	306

* The number of papers originating in five major industrial organizations (BTL, GE, Westinghouse, RCA, IBM) is given in parentheses.

E. *International Conference on Superconductivity, Colgate University, 1963*

	Universities	Industry	Gov't	Foreign
No of Organizations	34	29	15	50
No. of Attendees	129	116	32	67
No. of Papers	33	32	1	23

Eighty-two of the 129 attendees from 34 universities (USA) came from 10 of the universities. (Massachusetts Institute of Technology, 20; University of Maryland, 10; Rutgers University, Stanford University, University of Chicago, University of California at Berkeley, 7 each; University of Illinois, University of Pennsylvania, California Institute of Technology and University of California at San Diego, 6 each.)

Seventy-seven of the 116 attendees from 29 industrial organizations came from 5 organizations. (GE, 20; BTL, 19; Westinghouse, 14; RCA, 13; and IBM, 11.)

F. *Conference on the Physics of Type II Superconductors, Western Reserve University, 1964*

	Universities	Industry	Gov't	Foreign
No. of Organizations	24	19	15	26
No. of Attendees	62	65	29	46
No. of Papers	8	19	5	11

Twenty-nine of the 62 attendees from 24 universities (USA) came from 5 universities. (Rutgers University, 9; Massachusetts Institute of Technology, 8; Cornell University, Rice Institute, and Western Reserve University, 4 each.)

Thirty-nine of the 65 attendees from 19 industrial organizations came from 5 organizations (RCA, 10; BTL, 8; Westinghouse, 8; GE, 7; and IBM, 6.)

From the papers published in *Physical Review* and in *Physical Review Letters* and those presented at meetings of the American Physical Society in the period 1961–65, we conclude that about 50 per cent of the research and development in superconductivity was in universities, a little more than 40 per cent was in industrial ogranizations, and a little less than 10 per cent was in governmental laboratories.

Also, from the numbers of publications and of papers presented at meetings, from attendance at meetings, and from some data obtained in connection with a study of markets for superconducting materials and magnets, together with a dash of intuition, we conclude that there were in 1964, about 375 scientists of Ph.D. level active in this field, and that, for these, about 200 were in universities; about 140, in industrial organizations; and about 35, in governmental laboratories.

Appendix 3

Estimates of Expenditures on Superconductivity Research and Development in USA in 1964

The Pake Report (*Physics Today*, April, 1966) estimated that in 1963 the cost per Ph.D. physicist per year was about $36,000 in universities, about $57,000 in industrial laboratories, and about $83,000 in government laboratories.

For 1964 we assumed an increase of about 5 per cent in annual support, or cost, over 1963, for universities and for industrial laboratories; i.e., costs of about $37,500 per Ph.D. in universities and of $60,000 per Ph.D. in industry.

The figure quoted in the Pake Report for governmental laboratories, $83,000, is high because of the relatively lower fraction of staff in these laboratories with a Ph.D. (see, "Who are Physicists? What Do They Do?" in *Physics Today*, January, 1966) and because of the more expensive facilities utilized in many of the laboratories (especially AEC and NASA laboratories). We considered that in superconductivity research and development, costs per Ph.D. per year would approximate those in industry, i.e., about $60,000 per year.

Using the numbers of Ph.D.'s arrived at, 200 in universities, 140 in industry, and 35 in government laboratories, we estimated the annual expenditures as follows:

Universities 200 @ $37,000   7,400,000 $\simeq$ 7.5 $\times$ 10^6
Industries   140 @ $60,000   8,400,000 $\simeq$ 8.5 $\times$ 10^6
Government   35 @ $60,000   2,100,000 $\simeq$ 2.0 $\times$ 10^6
                    Total 18 $\times$ 10^6

This amount, $18 million for 1964, seems consistent with our estimate of $78 million for 1961–65, since annual expenditures were lower in the earlier years and probably higher in 1965.

Although this is not a very precise estimate, we believe the estimate of total expenditures and the estimates for expenditures in universities and industry are probably within plus or minus 25 per cent of the actual expenditures, and we suspect a somewhat wider margin of error in the governmental segment.

Appendix 4

Estimates of AEC Expenditures on Superconductivity Research and Development

We have recently received some data on AEC expenditures on research and development in the area of superconductivity from Louis C. Ianniello of the Division of Research of the Atomic Energy Commission, and he has given us permission to quote them. These figures were prepared at the request of the Joint Committee on Atomic Energy of the Congress and represent the AEC's best estimates for fiscal years 1964 through 1967.

	FY 1964	FY 1965	FY 1966	FY 1967
Expenditures ($000's)	1,653	2,554	3,085	4,284

The figures for FY 1966 and FY 1967 are only preliminary.

From the figures for FY 1964 and FY 1965, one may estimate the AEC's expenditures in *calendar year* 1964 as about $2.1 million.

Subject Index

Abrikosov's model for Type II superconductors, 12
Alternating-current transmission on superconducting lines: anelastic losses, 92, 96; hysteresis losses, 95
Anelastic losses in ac transmission, 82
Anisotropy of the gap via tunneling, 24
Armature of an alternator, 120
Au_5Ba, 52
Au_5Ca, 52

BCS theory, 2–7
Be, 56, 58
Balanced rectifier circuit, 125
Bardeen, Cooper, and Schrieffer, theory of. *See* BCS theory
Bloch states, 5
Bose-Einstein condensation, 9

C_8K, 53
Coherence length, 36; various material types, 54–55; very small, 53
Cold working techniques, 49
Computers, 104–5
Condensate, 1
Condensation of the electrons in momentum space, 6
Corresponding states, law of, 14
Coulomb interaction, 5
Coupling constant, N(o)V, 52
Critical fields: H_c, 4; H_{c1}, H_{c2}, H_{c3}, 36, 116–17; theoretical dependence on the Ginzburg-Landau parameter, 37
Critical magnetic field, 2; of a Type II superconductor, 4

Critical temperature: isotopic mass, 2; upper limit, 16; of a Pb-Sn composite, 46; for various material types, 54–55
Cryotron, 104–5
Current-carrying state, 9
Current density, relation to free energy, 11
Current problems, in superconductivity, 15–16
Current voltage characteristic, of a superconducting tunnel junction, 20, 22, 67; steps in, 27

dc transmission on superconducting lines, 81–91
Debye temperature, 52
Defects in materials, 49, 106–7
Density of states, 21, for electrons in a superconductor, 23, 52; in tunneling, 23
Dissipation: due to fluctuating demand, 81; caused by pure normal-metal solder joints, 82–83
Double-switch flux shifter, 125

Electron–phonon interaction, responsible for superconductivity, 3, 53
Energy, of a quasi-particle, 7
Energy gap, 1, 3, 7; anisotropy of, 24
Ettingshausen effect, 42
Excited states of a superconductor, 6–7

Fast neutron irradiation, 49
Fermi: level, 5; surface, 5

157

SUBJECT INDEX

Ferromagnetism and superconductivity, 15
Fluctuating demand, 81
Flux: magnetic, quantization of (*see* quantization of magnetic flux); flow resistivity, 14; magnetic, in a tunnel junction, 27; creep, 49; pumps, 104, principle of operation, 121–26; jumping, 111–12; spot voltage generator, 122–23
Frauenhoffer diffraction pattern, 27, 66
Free energy, in Ginzburg-Landau theory, 32
Fundamental constant h/e, measurement of, 28

Gapless superconductivity caused by paramagnetic impurities, 24
Geometrical effects in homogeneous superconducting material, 42–46
Ginzburg-Landau equation, derivation of, 31–32
Ginzburg-Landau parameter $\varkappa$, 33; and electron mean free path, 33, 53, 66
Ginzburg-Landau theory, 3, 31–33; Abrikosov's solution for Type II superconductors, 3
GLAG theory, 5

Hard superconductor. *See* Type II superconductors
Heat treatment techniques, 49
Heavy normal shunting, internal cooling approach, 113–15
High-current superconductivity, 56–57
High-field superconductors, 47–50; in relation with high normal resistivities, 33, 108; synthetic, 50
Hysteresis model, 119

Incomplete sintering techniques, 49
Inhomogeneous superconducting materials, 47–50
InSb, 56, 58
Interconnection of superconducting transmission line to room-temperature power network, 87
Intermediate state, domain structure of, 44–45
Isotope effect, 2–3

Josephson current, 4, 66
Josephson effect: under fields, 25–27; and plasma oscillations, 28

Josephson junction: microwave radiation, 28, 72; used as a superconducting interferometer theory, 69–70; experiment, 71–74
Josephson tunneling. *See* Tunneling of supercurrents

La-0.8 at % Gd, 56
Lead: critical temperature, 8; in relation to strong coupling theory and tunneling, 22–23; as a Type I(b) material, 36, 53
London: theory, 2–3; equations, 64
Losses: at microwave frequencies, 8; in flux flow, 41, 47; in flux creep, 49; in alternating current transmission, 95; anelastic, 96; in Type II materials, 117–19

Macroscopic quantum mechanics, 63
Materials: synthetic high-field superconductors, 50; commonly used in superconducting magnet construction, 107
Matthias regularities, 51
Mean free path of electrons, 46, 50
Meissner effect, 2
Metal vacuum interface, 32, 42
Microwave absorption, 8
Mixed state, 12, 107
MoIr, 56

Nb: filaments, 50; as an intrinsic Type II superconductor, 53
Nb_3Sn, 80; as an important supermagnet material, 53; in superconducting transmission lines, 93; properties of, 105–6
Nb+50% Ti, 107
Nb+25% Zr, 107
Nernst-Ettingshausen, effect 42
Normal shunts in superconducting magnets, 110, 113–15

Order parameter, in Ginzburg-Landau theory, 31

Pair density, 31
Paired states, probability of occupation of, 7
Pairing of electrons, 1; with antiparallel spin, 6; with momentum zero, 6
Peltier effect, 42
Penetration depth, 3
Perfect diamagnetism, model of superconductor, 3
Persistent currents, 2, 9, 104

SUBJECT INDEX

Phase: of superconductive wave function, 9, 63; of the superconductive wave function, in tunneling, 24-25; of wave function, gauge invariant, 66
Phases of a superconductor: definition, 35; normal, sheath, super, and vortex, 38-42
Phase transition: second order, 1; first order, 36; in ferromagnetic materials to a superconducting state, 57; in superconducting semimetal and degenerate semiconductor, 57
Phonons: role in producing superconductivity, 5-6, 23; effect on tunneling, 23
Point contacts, 73-74
Premature normalization in superconducting magnet, cause of, 111-12
Probability amplitude, for occupation of a pair. *See* Paired states
Proximity effects in heterogeneous structures, 46-47

Q values, for microwave cavities, 104
Quantization of magnetic flux, 2, 9-10
Quantum interferometer, 69-71
Quasi-particles in a superconductor, 2

Refrigeration system for superconducting transmission line: design, 83-87; calculations, 96-99; vacuum pumps, 99
Roberts survey, 51

Small ferromagnetic particles technique, 49
$SrTiO_3$, 53
Strong coupling superconductivity, effect on tunneling, 22
Superconducting cable, 80-82; choice of material, 80; choice of transmission voltage, 80; design, 82
Superconducting devices: microwave detectors, 28; interferometers, 70-71; magnetometers, 74-75; digital fluxmeter, 76; high-Q microwave cavities, 104; high-Q circuits, 104; persistent current inductor, 104; switching devices, 104; cryotron, 104-5, 147; flux pumps, 104, 121-26, 147; and Type II superconductors, 105-16; size of, 120; high-field rotor magnet, 120-21; magnets, 120-21; power transformer, 121, 148; transmission lines, 77-101, 120, 148; gyroscopes, 147; linear accelerators, 147; bolometers, 148; generators, 148; lenses, 149

Superconducting ground state, 6
Superconducting magnet design, calculations, 107
Superconducting semiconductors, 52-53
Superconducting transmission lines, 79-92; protections against power interruption, safety of, maintenance of, 88-89; cost estimate of, 89-91; summary of design characteristics of, 91; problem of differential expansion in, 99-100
Superconductivity: occurrence of, 50-56; in one-dimensional systems, 57; economic aspects of, 127-50; research and developments of, 130; publication on, 131-32; expenditure on, 131, 150-51, 154; government incentives, 134-39; industrial incentives, 139-42; university incentives, 142-44
Supercooling, of the normal phase, 37
Supercurrent, flow, 1, 8-11, 20
Surface resistance, 8
Surface superconductivity, 57

Technical methods in compound magnet conductor, 107
Techniques for establishing trapping centers in supermagnet materials, 49
Th-rich matrix, 50
Thin films, 44
Thin film techniques, 19
Ti-16 at % Mo disordered solid solution alloy, 53
Transition temperature, 52
Tungsten as a Type I superconductor, 53
Tungsten bronzes, $Ba_{0.13}WO_3$, 56
Tunneling current, 21; formulas for, 21, 25, 66
Tunneling density of states, 23, 47
Tunneling of a supercurrent, 4, 24-28, 65; quenching by a magnetic field, 26-27, 66; measurement of the fundamental constant h/e, 28
Tunneling of electrons, 19-29
Twin-duct jacket, 81
Type I superconductor, 3, 93; Type I(a) and Type I(b), definition, 35
Type II superconductors, 12-15, 93, 104-5, 116-17; vortex model, 12, 36; definition, 35, 64; critical parameters of, 105-7

Velocity of propagation of phase boundary in magnet winding, 109

159

Vortex lines, 4; low energy excitations in the core of, 12, 15; neutron diffraction by, 12, 36; motions, 12–15, 41–42, 81, 94

Vycor glass, 50

Wave function of BCS, 5
Weak links, 73–74

Name Index

Abrikosov, A. A., 2, 4, 5, 11, 12, 16, 24, 29, 35, 58, 59
Alden, T. H., 60
Ambler, E., 61
Anderson, P. W., 2, 16, 26, 29, 60, 66, 76
Andres, K., 61
Arrhenius, G., 61
Atherton, D. L., 58, 62

Bardeen, J., 1, 2, 3, 16, 17, 58, 59
Barnes, L. J. 59
Bean, C. P., 50, 59, 60, 90, 95, 101, 116, 150
Benz, M. G., 100
Berlincourt, T. G., 31, 108
Bierstedt, P. E., 61
Biondi, M. A., 7, 8, 17
Bither, T. A., 61
Blatt, J. M., 60
Böbel, G., 60
Boato, G., 59
Bömmel, H. E., 62
Bon Mardion, G., 59
Bozowski, S., 62
Broadston, S., 61
Buchhold, T. A., 125
Buck, D. A., 143

Cardona, M., 14, 59, 61
Caroli, C., 12, 17, 59
Cline, H. E., 60, 100
Cody, G. D., 53, 61
Cohen, M. L. 52, 53, 61
Colwell, J. H., 61
Compton, V. B., 50, 60

Cooper, L. N., 2, 3, 16, 58
Cope, J. A., 59
Corenzwit, E., 50, 60, 61
Cribier, D., 17, 58
Cyrot, M., 59

Darnell, A. J., 62
Darnell, F. J., 61
Daunt, J. G., 52, 61
Deaver, B. S., Jr., 2, 10, 16
DeBlois, R. W., 59
DeGennes, P. G., 4, 12, 16, 17, 35, 57, 58, 59
DeSorbo, W., 59
D'Heurle, F. M., 100
Doll, R., 2, 10, 16
Dorer, G. L., 62
Druyvestyn, W. F., 59

Eagles, D. M., 57, 62
Essmann, U., 58, 62

Fairbank, W. M., 2, 10, 16
Falk, D. S., 60
Farnoux, B., 17, 58
Fawcett, D., 59
Felici, G., 125
Fink, H. J., 59
Finnemore, D. K., 56, 61, 62
Fiory, A. T., 15, 17, 59
Fleischer, R. L., 60, 101
Fontanella, E., 150, 151
Forgacs, R. L., 75, 76
Frederikse, H. P. R., 61
Fröhlich, H., 2, 3, 16, 57, 62

161

NAME INDEX

Gallinaro, G., 59
Garfinkel, M. P., 7, 8, 17
Garland, J. W., 52, 61
Garwin, R. L., 77, 120
Gatos, H. C., 62
Geballe, T. H., 50, 60, 61
Geller, S., 62
Giaever, I., 2, 4, 15, 16, 17, 19, 22, 28, 143
Gibson, J. W., 61
Giessen, B. C., 62
Ginzburg, V. L., 1, 4, 5, 16, 25, 32, 57, 58, 62
Gittleman, J. I., 62
Glover, R. E., 62
Goodman, B. B., 2, 16, 59
Gor'kov, L. P., 4, 5, 16, 24, 29, 33, 58
Gorter, C. J., 16
Graham, G. M., 62
Gygax, S., 61

Hake, R. R., 58, 59
Hamilton, D. C., 61
Hart, H. R., 59, 60, 101
Hartlin, E. M., 62
Hauser, J. J., 49, 60
Hein, R. A., 61
Helfand, E., 59
Heller, W. R., 100
Hempstead, C. F., 17, 59, 60
Hill, H. H., 62
Hohenberg, P. C., 59
Hopkins, D. C., 56, 62
Hosler, W. R., 61
Howard, J. G., 78, 100
Hull, G. W., 50, 60, 61, 62
Hulm, J. K., 61, 103
Hannay, N. B., 61

Ianniello, L. C., 155

Jaccarino, V., 57, 62
Jacrot, B., 17, 58
Jaklevic, R. C., 76
Jensen, M. A., 61
Johnson, R. T., 61
Joseph, A. S., 59, 60, 62
Josephson, B. D., 1, 2, 4, 16, 19, 28, 60, 65, 66, 76

Kantrowitz, A. R., 100
Kapitza, P., 50, 60
Kessinger, R. D., 59
Kim, Y. B., 12, 14, 15, 17, 59, 60
Kirzhnits, D. A., 57, 62
Kresin, W. S., 60
Kunzler, J. E., 93, 94, 105, 143

Lacaze, A., 59
Lambe, J., 76
Landau, L. D., 1, 4, 5, 16, 25, 32, 58
Langenberg, D. N., 28
Laverick, C., 100, 113
Lazarev, G. G., 62
Libby, W. F., 62
Little, W. A., 57, 62
Livingston, J. D., 49, 59, 60
London, F., 1, 2, 3, 5, 16, 63, 76, 116
Lynton, E. A., 60

McFee, R., 100
MacNair, D., 61
McWhan, D. B., 62
Maki, K., 59, 60
Marcus, P. M., 59
Matisoo, J., 77
Matricon, J., 12, 17
Matthias, B. T., 17, 50, 51, 60, 61, 62
Maxwell, E., 2, 16, 25
Mazelsky, R., 61
Meissner, H., 94
Meissner, W., 2, 3, 4, 16
Mercereau, J. E., 29, 63, 76
Merriam, M. F., 61
Miller, P. B., 7, 8, 17
Miller, R. C., 61
Minnigerode, G. v., 49, 60
Montgomery, D. B., 100
Morrison, W. A., 127
Mydosh, J., 94

Näbauer, M., 2, 10, 16
Neissen, A. K., 15, 17, 59
Nesbitt, L. B., 16

Ochsenfeld, R., 2, 3, 16
Olmsted, L. M., 78, 100
Olsen, C. E., 62, 125
Onnes, H. Kamerlingh, 2, 16, 143
Otter, F. A., 15, 17, 59, 88

Palmer, P. E., 56, 62
Park, J. G., 59
Parker, W. H., 28
Parmenter, R. H., 56, 62
Peter, M., 57, 62
Pfeiffer, E. R., 61
Pippard, A. B., 144
Pollard, E. R., 62

Rao, L. M., 17, 58
Ratto, C. I., 60
Raub, Ch. J., 61
Reed, W. A., 59
Reif, F., 24, 29, 60
Reynolds, C. A., 2, 16

162

NAME INDEX

Rizzuto, C., 59
Roberts, B. W., 50, 51, 60, 61
Rose, R. M., 60
Rosenblum, B., 14, 59, 61, 62
Rosi, F. D., 62
Rowell, J. M., 4, 16, 26, 29, 60, 66, 76

Sadagopan, V., 62
Saint-James, D., 4, 16, 35, 57, 58
Sampson, W., 100
Sandiford, D. J., 59
Scalapino, D. J., 23, 28
Schaafs, W., 60
Schadler, H. W., 49, 60
Schmitt, P., 61
Schmitt, R. W., 127
Schooley, J. F., 61
Schrieffer, J. R., 2, 3, 16, 23, 28, 58
Schubnikov, L. V., 4, 16
Schweitzer, D. G., 59
Scott, R. B., 100, 101
Seidel, T. E., 62
Selivanenko, A. S., 57, 62
Semenenko, E. E., 62
Seraphim, D. P., 100
Serin, B., 15, 16, 17, 59
Silver, A. H., 76
Silvert, W., 57, 62
Singh, A. D., 60
Smithells, C. J., 101
Solomon, P. R., 15, 17, 59
Spreadbury, F. G., 100
Staas, F. A., 59
Stekly, Z. J. J., 100, 113
Stephen, M. J., 15, 17, 59
Strauss, B. P., 60
Strnad, A. R., 17, 59, 60

Stromberg, T. F., 61, 62
Sudovtsov, A. I., 62
Swartz, P. S., 59, 60, 101, 150
Sweedler, A. R., 61
Swenson, C. A., 61, 62

Tavger, B. A., 60
Taylor, B. N., 28
Terreaux, C., 57, 62
Theurer, H. C., 49, 60
Thompson, C. J., 60
Thurber, W. R., 61
Thus, W., 58
Tinkham, M., 60
Tittman, B. R., 62
Tomasch, W. J., 58, 60
Trauble, H., 58, 62

Van Gurp, G. J., 15, 17
Van Ooyen, D. J., 15, 17
Van Viejfeijken, A. G., 15, 17
Vilches, O. E., 61
Volger, J. A., 122, 123

Warnick, A., 75, 76
Werthamer, N. R., 46, 59, 60
Wertheimer, R. M., 62
Wheatley, J. C., 61
Wicklund, A. W., 62
Wilkins, J. W., 23, 28
Wilkinson, K. J. R., 101
Wipf, S. L., 122, 123, 124
Woolf, M. A., 24, 29, 60
Wright, W. H., 16
Wulff, J., 60

Zavaritskii, N. V., 23, 24, 29